D1461851
CGP
CGP
– books
like no others!

Get your facts straight with CGP!

This CGP Knowledge Organiser has one mission in life — helping you remember the key facts for AQA GCSE Biology.

We've boiled every topic down to the vital definitions, facts and diagrams, making it all easy to memorise.

There's also a matching Knowledge Retriever book that'll test you on every page. Perfect for making sure you know it all!

CGP — still the best! ☺

Our sole aim here at CGP is to produce the highest quality books — carefully written, immaculately presented and dangerously close to being funny.

Then we work our socks off to get them out to you — at the cheapest possible prices.

Contents

Working Scientifically

The Scientific Method 2
Designing & Performing Experiments 3
Presenting Data 4
Conclusions, Evaluations and Units 5

Topic 1 — Cell Biology

Cells 6
Cell Division 7
Cell Specialisation and Stem Cells 8
Transport in Cells 9
Exchanging Substances 10

Topic 2 — Organisation

Cell Organisation and Enzymes 11
The Lungs and the Heart 12
Blood Vessels and Blood 13
Cardiovascular Disease 14
Health and Disease 15
Risk Factors for Diseases and Cancer 16
Plant Cell Organisation 17
Transpiration 18

Topic 3 — Infection and Response

Communicable Disease 19
Fighting Disease 20
Drugs 21
Monoclonal Antibodies 22
Plant Diseases and Defences 23

Topic 4 — Bioenergetics

Photosynthesis 24
Respiration 25
Metabolism and Exercise 26

Topic 5 — Homeostasis and Response

Homeostasis and the Nervous System 27
Synapses, Reflexes and the Brain 28
The Eye 29
Body Temperature and Hormones 30
Blood Glucose and Hormones 31
Waste Substances and the Kidneys 32
Puberty and the Menstrual Cycle 33
Controlling Fertility 34
Plant Hormones 35

Topic 6 — Inheritance, Variation and Evolution

DNA 36
DNA, Proteins and Mutations 37
Reproduction 38
Genetic Diagrams 39
Inherited Disorders 40
Developments in Genetics & Variation 41
Evolution and Speciation 42
Uses of Genetics 43
Fossils and Antibiotic Resistance 44
Classification and Extinction 45

Topic 7 — Ecology

Basics of Ecology 46
Food Chains & Environmental Change 47
Cycling of Materials 48
Decay and Biodiversity 49
Human Effects on Ecosystems 50
Maintaining Ecosystems 51
Trophic Levels 52
Food Security and Biotechnology 53

Required Practicals

Required Practicals 1 54
Required Practicals 2 55
Required Practicals 3 56
Required Practicals 4 57
Required Practicals 5 58
Required Practicals 6 59

Practical Skills

Measuring and Sampling 60
Practical Techniques 61
Techniques and Results 62

Published by CGP.
From original material by Richard Parsons.

Editors: Luke Bennett, Ellen Burton, Sam Mann, Rachael Rogers.
Contributors: Susan Alexander and Paddy Gannon.

With thanks to Susan Alexander for the proofreading.
With thanks to Emily Smith for the copyright research.

ISBN: 978 1 78908 488 7

Printed by Elanders Ltd, Newcastle upon Tyne.
Clipart from Corel®
Illustrations by: Sandy Gardner Artist, email sandy@sandygardner.co.uk

The Scientific Method

Developing Theories

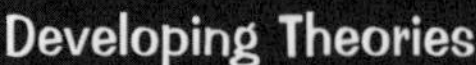

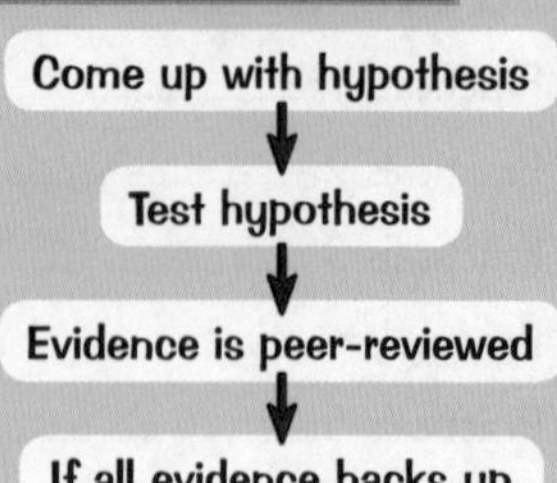
Come up with hypothesis
Test hypothesis
Evidence is peer-reviewed
If all evidence backs up hypothesis, it becomes an accepted theory.

HYPOTHESIS — a possible explanation for an observation.

PEER REVIEW — when other scientists check results and explanations before they're published.

Accepted theories can still change over time as more evidence is found, e.g. inheritance and genetics:

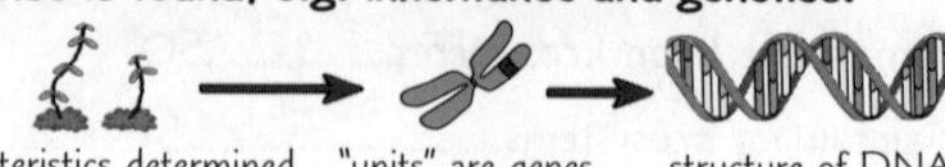

characteristics determined by "hereditary units" → "units" are genes on chromosomes → structure of DNA shows how genes work

Models

REPRESENTATIONAL MODELS — a simplified description or picture of the real system, e.g. lock and key model of enzyme action:

enzyme
active site
substrate
fit like lock and key

Models help scientists explain observations and make predictions.

COMPUTATIONAL MODELS — computers are used to simulate complex processes.

Issues in Science

Scientific developments can create four issues:

1. **Economic** — e.g. beneficial technology, like alternative energy sources, may be too expensive to use.
2. **Environmental** — e.g. new technology could harm the natural environment.
3. **Social** — decisions based on research can affect society, e.g. taxing alcohol to reduce health problems.
4. **Personal** — some decisions affect individuals, e.g. a person may not want a wind farm being built near to their home.

Media reports on scientific developments may be oversimplified, inaccurate or biased.

Hazard and Risk

HAZARD — something that could potentially cause harm.

RISK — the chance that a hazard will cause harm.

Hazards associated with biology experiments include:

Some microorganisms can make you ill.

Hazardous chemicals (e.g. flammable ethanol).

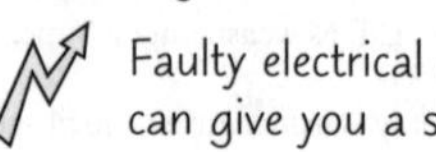

Faulty electrical equipment can give you a shock.

The seriousness of the harm and the likelihood of it happening both need consideration.

Designing & Performing Experiments

Collecting Data

Data should be...	
REPEATABLE	Same person gets same results after repeating experiment using the same method and equipment.
REPRODUCIBLE	Similar results can be achieved by someone else, or by using a different method or piece of equipment.
ACCURATE	Results are close to the true answer.
PRECISE	All data is close to the mean.

Reliable data is repeatable and reproducible.

Valid results are repeatable and reproducible and answer the original question.

Fair Tests

INDEPENDENT VARIABLE	Variable that you change.
DEPENDENT VARIABLE	Variable that is measured.
CONTROL VARIABLE	Variable that is kept the same.
CONTROL EXPERIMENT	An experiment kept under the same conditions as the rest of the investigation without anything being done to it.
FAIR TEST	An experiment where only the independent variable changes, whilst all other variables are kept the same.

An "indie" variable

Control experiments are carried out when variables can't be controlled.

Four Things to Look Out For

1. **RANDOM ERRORS** — unpredictable differences caused by things like human errors in measuring.
2. **SYSTEMATIC ERRORS** — measurements that are wrong by the same amount each time.
3. **ZERO ERRORS** — systematic errors that are caused by using a piece of equipment that isn't zeroed properly.
4. **ANOMALOUS RESULTS** — results that don't fit with the rest of the data.

Anomalous results can be ignored if you know what caused them.

Processing Data

Calculate the mean — add together all data values and divide by number of values.

UNCERTAINTY — the amount by which a given result may differ from the true value.

$$\text{uncertainty} = \frac{\text{range}}{2}$$

range: largest value minus smallest value

In any calculation, you should round the answer to the lowest number of significant figures (s.f.) given.

Presenting Data

Bar Charts

Bar charts are used when independent variable is categoric or discrete.

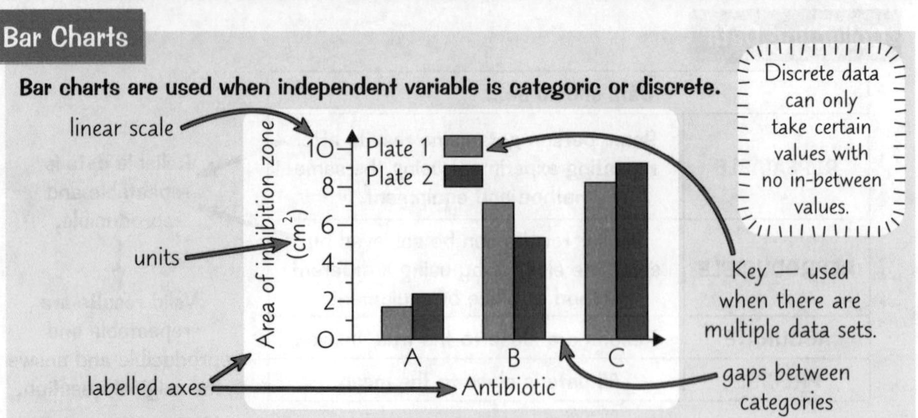

Discrete data can only take certain values with no in-between values.
linear scale
units
labelled axes
Area of inhibition zone (cm²)
10
8
6
4
2
0
Plate 1
Plate 2
A
B
C
Antibiotic
Key — used when there are multiple data sets.
gaps between categories

Plotting Graphs

Graphs are used when both variables are continuous.

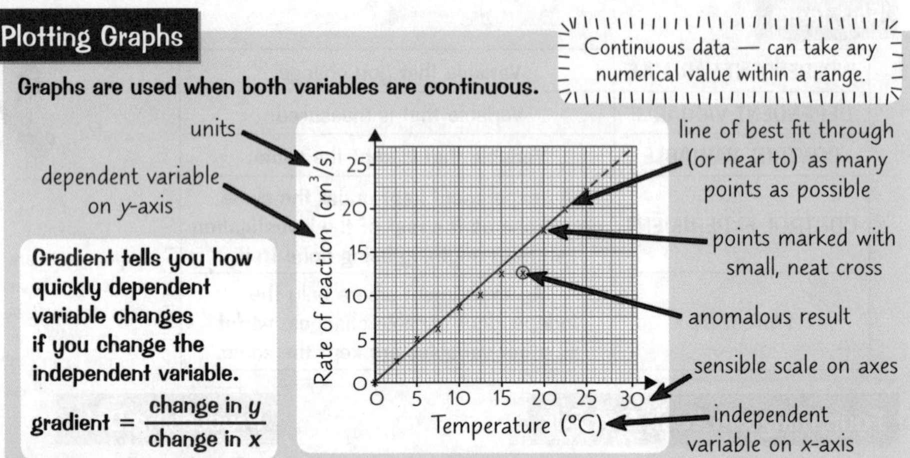

Continuous data — can take any numerical value within a range.
units
dependent variable on y-axis
Rate of reaction (cm³/s)
25
20
15
10
5
0
0 5 10 15 20 25 30
Temperature (°C)
line of best fit through (or near to) as many points as possible
points marked with small, neat cross
anomalous result
sensible scale on axes
independent variable on x-axis

Gradient tells you how quickly dependent variable changes if you change the independent variable.

$$\textbf{gradient} = \frac{\textbf{change in } y}{\textbf{change in } x}$$

Three Types of Correlation Between Variables

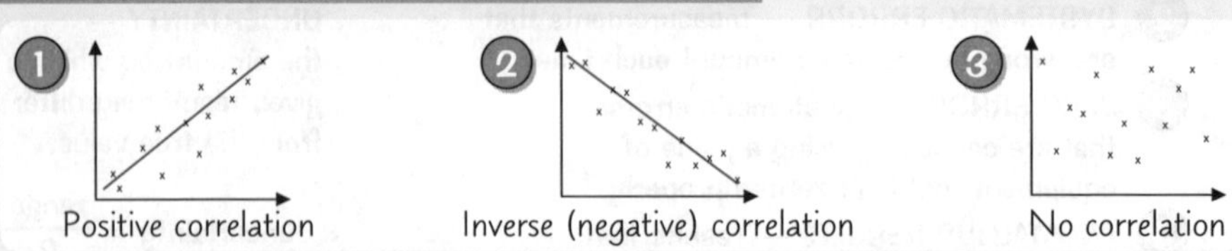

1
Positive correlation
2
Inverse (negative) correlation
3
No correlation

Possible reasons for a correlation:

Chance — **correlation might be a fluke.**

Third variable — **another factor links the two variables.**

Cause — **if every other variable that could affect the result is controlled, you can conclude that changing one variable causes the change in the other.**

Conclusions, Evaluations and Units

Conclusions

Draw conclusion by stating relationship between dependent and independent variables.

↓

Justify conclusion using specific data.

↓

Refer to original hypothesis and state whether data supports it.

You can only draw a conclusion from what your data shows — you can't go any further than that.

Evaluations

EVALUATION — a critical analysis of the whole investigation.

	Things to consider
Method	• Validity of method • Control of variables
Results	• Reliability, accuracy, precision and reproducibility of results • Number of measurements taken • Level of uncertainty in the results
Anomalous results	• Causes of any anomalous results

Repeating experiment with changes to improve the quality of results will give you more confidence in your conclusions.

You could make more predictions based on your conclusion, which you could test in future experiments.

S.I. Units

S.I. BASE UNITS — a set of standard units that all scientists use.

Quantity	S.I. Unit
mass	kilogram (kg)
length	metre (m)
time	second (s)

Scaling Units

SCALING PREFIX — a word or symbol that goes before a unit to indicate a multiplying factor.

Multiple of unit	Prefix
10^{12}	tera (T)
10^{9}	giga (G)
10^{6}	mega (M)
1000	kilo (k)
0.1	deci (d)
0.01	centi (c)
0.001	milli (m)
10^{-6}	micro (μ)
10^{-9}	nano (n)

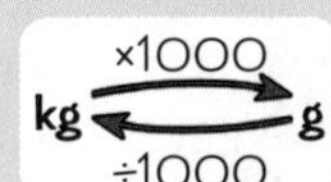
×1000
kg
g
÷1000

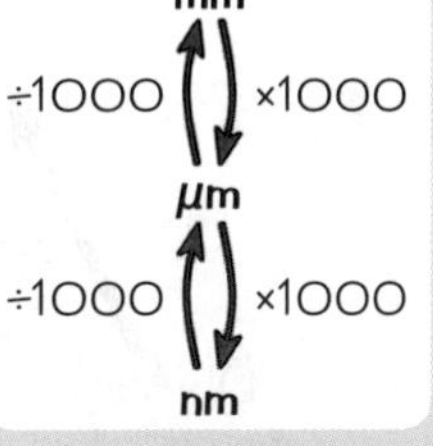
mm
÷1000
×1000
μm
÷1000
×1000
nm

Cells

Eukaryotic Cells

ANIMAL CELL

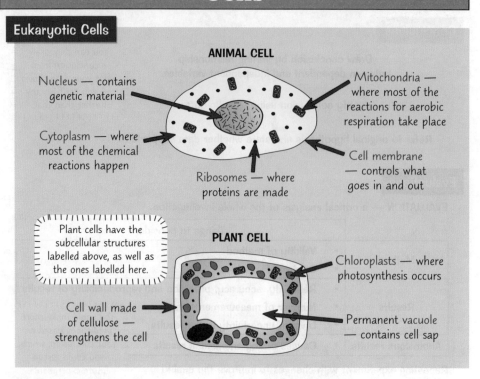

Nucleus — contains genetic material

Mitochondria — where most of the reactions for aerobic respiration take place

Cytoplasm — where most of the chemical reactions happen

Cell membrane — controls what goes in and out

Ribosomes — where proteins are made

Plant cells have the subcellular structures labelled above, as well as the ones labelled here.

PLANT CELL

Chloroplasts — where photosynthesis occurs

Cell wall made of cellulose — strengthens the cell

Permanent vacuole — contains cell sap

Prokaryotic Cells

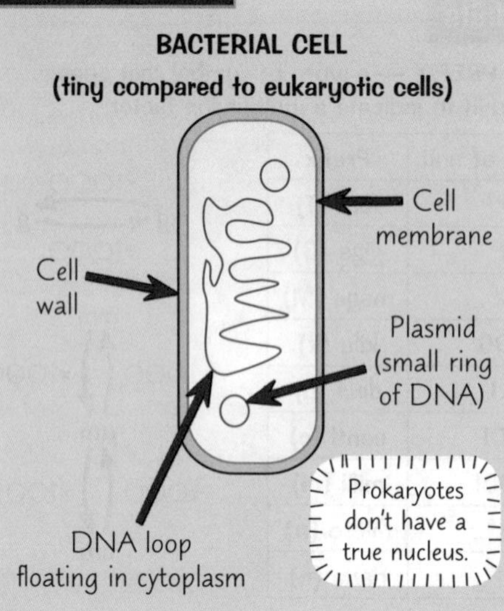
BACTERIAL CELL
(tiny compared to eukaryotic cells)
Cell membrane
Cell wall
Plasmid (small ring of DNA)
DNA loop floating in cytoplasm

Prokaryotes don't have a true nucleus.

Microscopy

Electron microscopes were invented later than light microscopes.

They have a higher magnification and resolution than light microscopes — they let us see smaller things in more detail, meaning we can understand subcellular structures better now.

$$\text{magnification} = \frac{\text{image size}}{\text{real size}}$$

Use standard form to write really small numbers. E.g. $0.0045 = 4.5 \times 10^{-3}$

Cell Division

Chromosomes and the Cell Cycle

CHROMOSOMES — coiled up lengths of DNA molecules, which carry genes. They're found in the nucleus, and they're normally in pairs in body cells.

CELL CYCLE — a series of stages in which cells divide to produce new cells.

Before a cell divides, it does three things:

1. **Grows in size.**
2. **Increases the amount of subcellular structures, e.g. mitochondria and ribosomes.**
3. **Duplicates its DNA.**

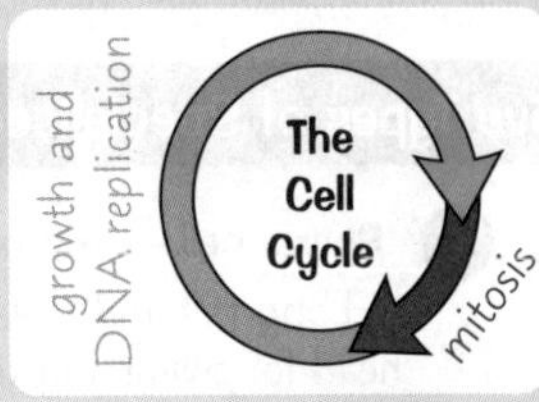
growth and DNA replication
The Cell Cycle
mitosis

Mitosis is okay, but check out my toesies.

Mitosis

MITOSIS — the stage of the cell cycle when the cell divides.

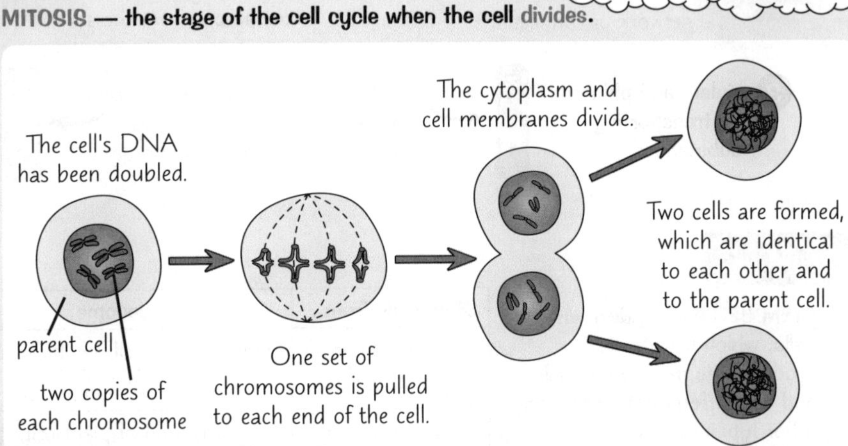
The cell's DNA has been doubled.
parent cell
two copies of each chromosome
One set of chromosomes is pulled to each end of the cell.
The cytoplasm and cell membranes divide.
Two cells are formed, which are identical to each other and to the parent cell.

Mitosis allows multicellular organisms to grow or replace cells that have been damaged.

The average time taken for a bacterial cell to divide in two is the 'mean division time'.

Binary Fission

BINARY FISSION — the simple cell division process by which bacterial cells replicate. It can happen as often as every 20 minutes if there are enough nutrients and the temperature is suitable.

DNA loop and plasmid(s) replicate

two daughter cells produced

Cell Specialisation and Stem Cells

Cell Differentiation

DIFFERENTIATION — the process by which a cell changes to become specialised for its job.

In most animal cells, the ability to differentiate is lost at an early stage, but many plant cells never lose this ability.

Five Types of Specialised Cells

1 Sperm cell — reproduction

Long tail and streamlined head for swimming to the egg.

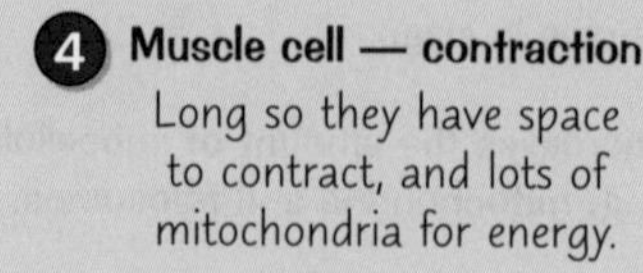

4 Muscle cell — contraction

Long so they have space to contract, and lots of mitochondria for energy.

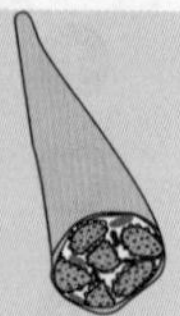

2 Nerve cell — rapid signalling

Long to cover a large distance, and branched to form a network of connections.

5 Root hair cell — absorbing water and minerals

Large surface area for absorbing water and mineral ions from the soil.

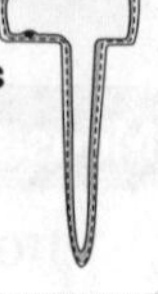

3 Xylem and phloem — transporting substances

Xylem cells are hollow and phloem cells have few subcellular structures, so substances can easily flow through.

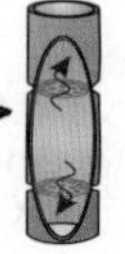

Stem Cells

STEM CELLS — undifferentiated cells, which can divide to produce lots more stem cells, and can differentiate into many other types of cell.

Stem cells from...	Can become...
adult bone marrow	many kinds of cell, e.g. blood cells
human embryo	any kind of human cell
plant meristem	any kind of plant cell

Stem cells can be grown in a lab and made to differentiate into specialised cells:

Uses in medicine	Uses in plants
E.g. stem cells could produce nerve cells to treat paralysis, or insulin-producing cells to treat diabetes. **In therapeutic cloning, an embryo could be made with the same genes as the patient — then stem cells used from the embryo wouldn't be rejected by the patient.**	**Produce clones of whole plants quickly and cheaply, e.g. to grow more plants of a rare species, or clone crops with desired features.**

Risk: stem cells from the lab could get a virus, which could get transferred to the patient.

Transport in Cells

Diffusion

DIFFUSION — the spreading out of particles from an area of higher concentration to an area of lower concentration.

Only very small molecules (e.g. oxygen, glucose) can diffuse through cell membranes.

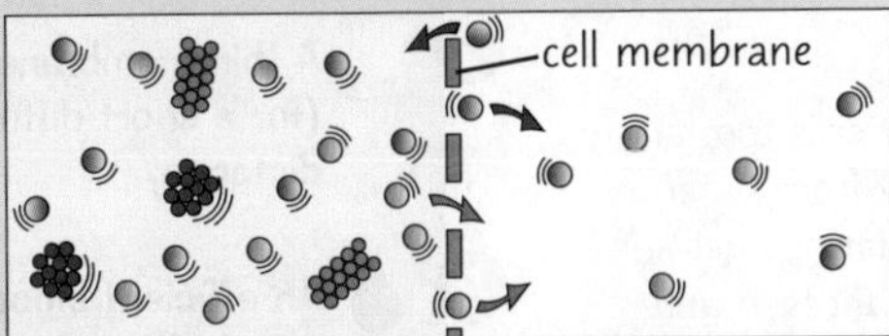
cell membrane

These three factors increase the rate of diffusion across a cell membrane:

1. 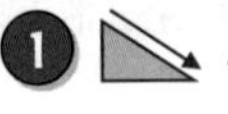 **A high concentration gradient** (e.g. loads of the particles on one side and hardly any on the other).
2.  **A high temperature.**
3.  **A large surface area.**

Osmosis

OSMOSIS — the movement of water molecules across a partially permeable membrane from a region of higher water concentration to a region of lower water concentration.

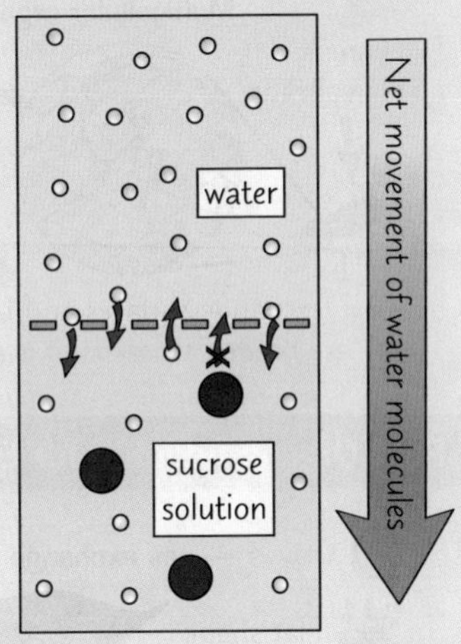
water
sucrose solution
Net movement of water molecules

Transport this info into your head, even if it's going against your concentration gradient.

Active Transport

ACTIVE TRANSPORT — the movement of a substance against the concentration gradient. Unlike diffusion and osmosis, it requires energy from respiration. It allows...

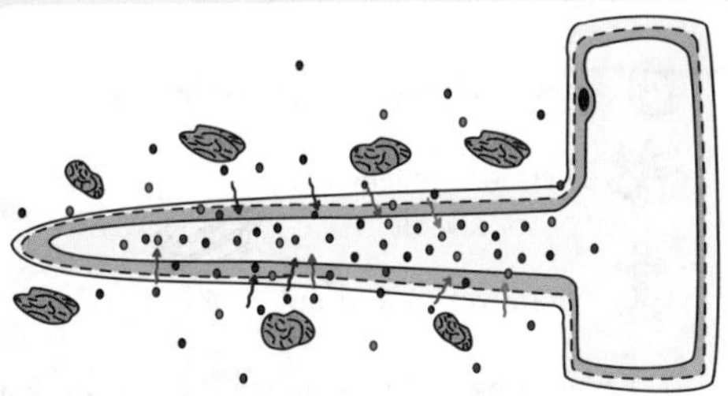

...mineral ions (for plant growth) to be absorbed from the soil into root hair cells.

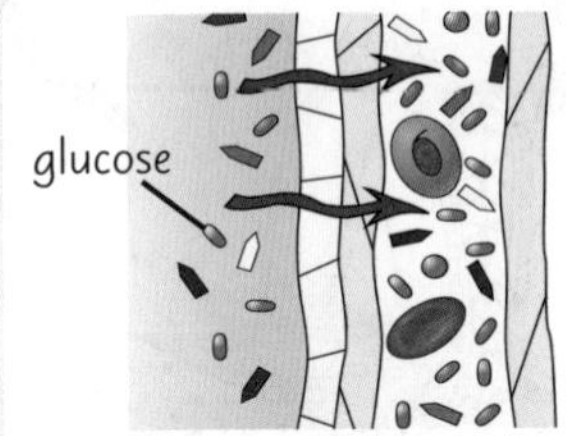
glucose

...glucose (for cell respiration) to be absorbed into the bloodstream from the gut.

Exchanging Substances

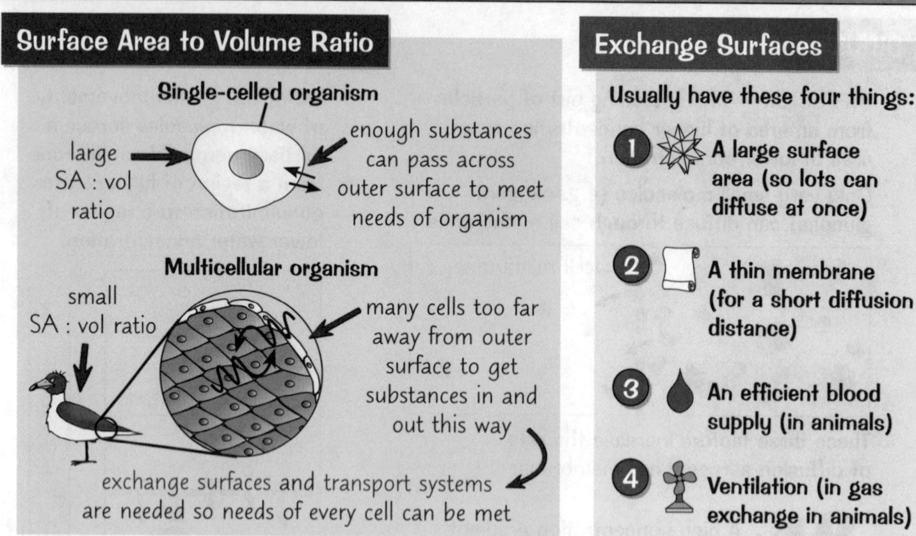
Surface Area to Volume Ratio
Single-celled organism
large SA : vol ratio
enough substances can pass across outer surface to meet needs of organism
Multicellular organism
small SA : vol ratio
many cells too far away from outer surface to get substances in and out this way
exchange surfaces and transport systems are needed so needs of every cell can be met
Exchange Surfaces
Usually have these four things:
1 A large surface area (so lots can diffuse at once)
2 A thin membrane (for a short diffusion distance)
3 An efficient blood supply (in animals)
4 Ventilation (in gas exchange in animals)

Four Organs Adapted for Exchange

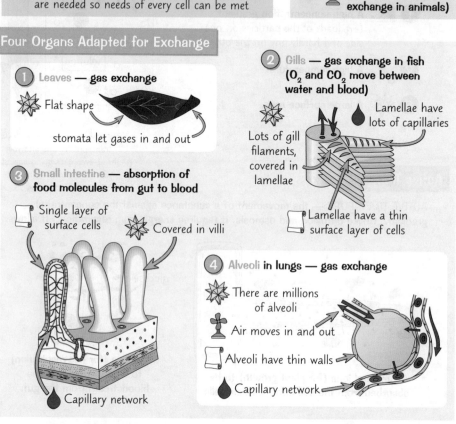

1 Leaves — gas exchange

Flat shape

stomata let gases in and out

2 Gills — gas exchange in fish (O_2 and CO_2 move between water and blood)

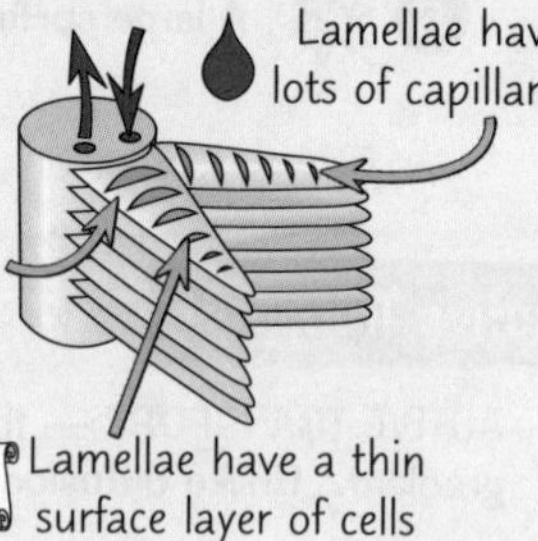

Lots of gill filaments, covered in lamellae

Lamellae have lots of capillaries

Lamellae have a thin surface layer of cells

3 Small intestine — absorption of food molecules from gut to blood

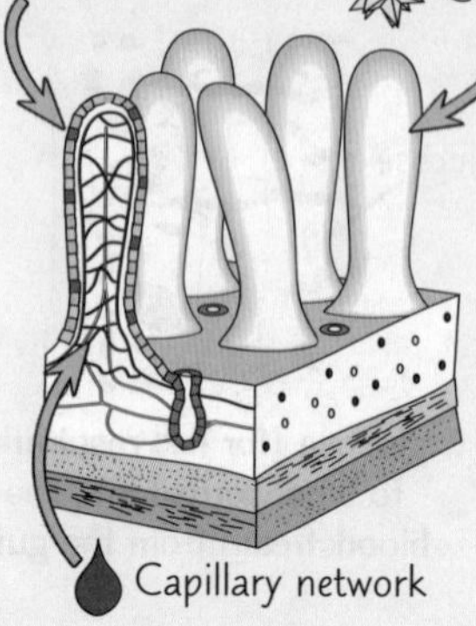

Single layer of surface cells

Covered in villi

Capillary network

4 Alveoli in lungs — gas exchange

There are millions of alveoli

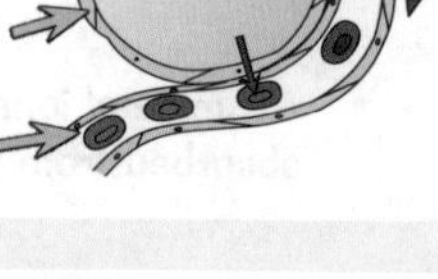

Air moves in and out

Alveoli have thin walls

Capillary network

Cell Organisation and Enzymes

Cell Organisation

CELL — a basic building block that all living organisms have.

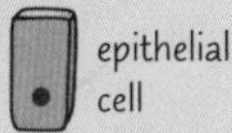
epithelial cell

TISSUE — a group of similar cells that work together.

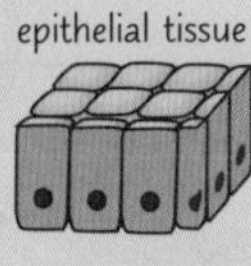
epithelial tissue

ORGAN — a group of different tissues that work together.

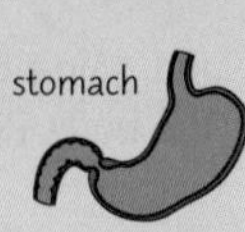
stomach

ORGAN SYSTEM — a group of organs working together.

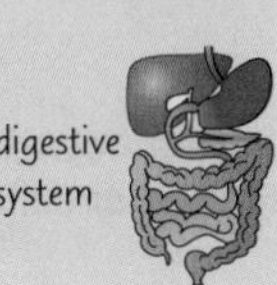
digestive system

Organ systems work together to make entire organisms.

Enzymes

Enzymes catalyse (speed up) chemical reactions.

Each enzyme only catalyses one specific reaction because of the unique shape of its active site.

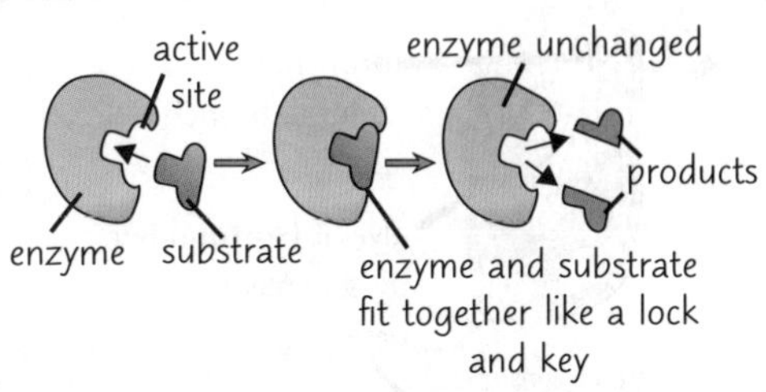
active site
enzyme unchanged
products
enzyme
substrate
enzyme and substrate fit together like a lock and key

High temperatures and high and low pHs change the shape of the active site so the enzyme no longer works.

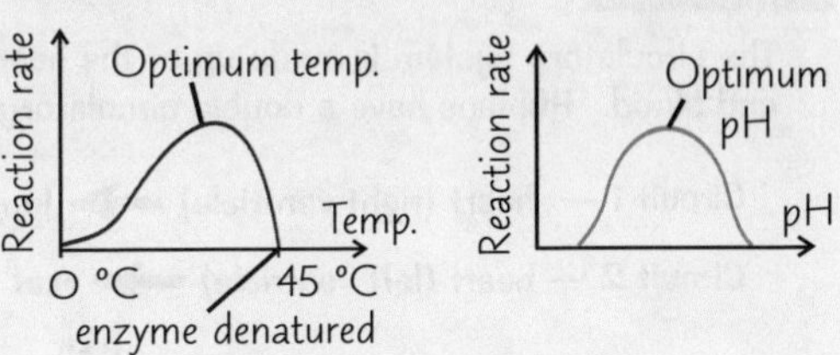
Reaction rate
Optimum temp.
Temp.
0 °C
45 °C
enzyme denatured
Reaction rate
Optimum pH
pH

Digestion

Digestive enzymes break BIG molecules down into smaller, soluble ones. These can pass through the walls of the digestive system and be absorbed into the blood.

Enzyme	Breaks down...	Into...	Produced in the...
amylase (a carbohydrase)	**starch** (a carbohydrate)	**maltose and other sugars**	**salivary glands, small intestine, pancreas**
protease	**protein**	**amino acids**	**stomach, small intestine, pancreas**
lipase	**lipid**	**glycerol and fatty acids**	**small intestine, pancreas**

The products of digestion can be used to make new carbohydrates, proteins and lipids.

Bile speeds up digestion in two ways:

 **It makes conditions alkaline so enzymes in the small intestine work better.**

 **It emulsifies fat so there's a larger surface area for lipase to work on.**

Bile is made in the liver...

...and stored in the gall bladder.

The Lungs and the Heart

The Lungs

Gas Exchange

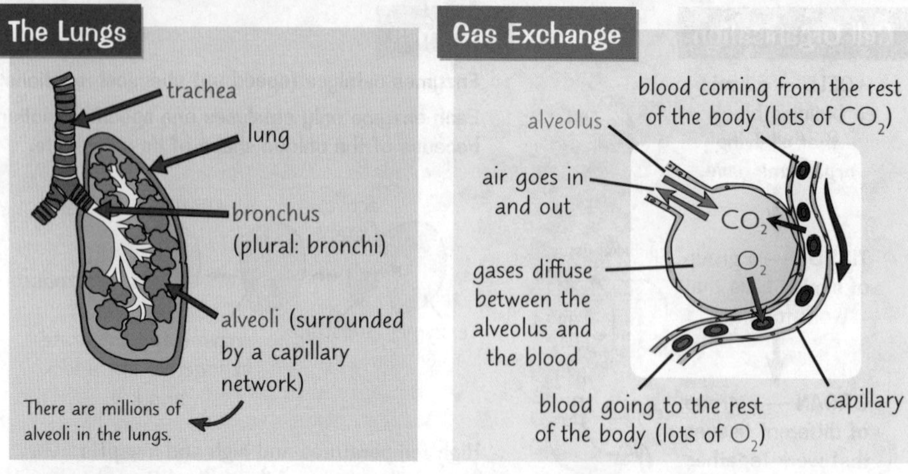
trachea
lung
bronchus
(plural: bronchi)
alveoli (surrounded
by a capillary
network)
There are millions of
alveoli in the lungs.
alveolus
blood coming from the rest
of the body (lots of CO_2)
air goes in
and out
CO_2
O_2
gases diffuse
between the
alveolus and
the blood
blood going to the rest
of the body (lots of O_2)
capillary

The Heart

The circulatory system is made up of the heart (a pumping organ), blood vessels and blood. Humans have a double circulatory system (two circuits):

Circuit 1 — heart (right ventricle) → lungs → heart

Circuit 2 — heart (left ventricle) → rest of body → heart

KEY:
= oxygenated blood
= deoxygenated blood

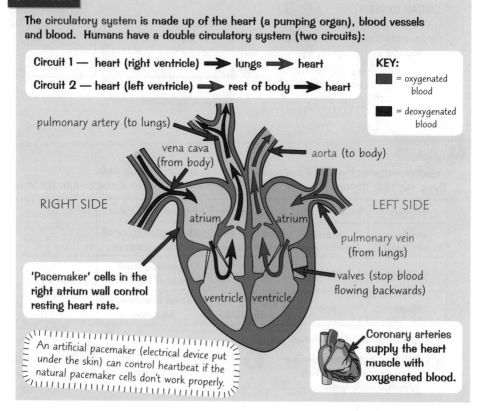

pulmonary artery (to lungs)
vena cava (from body)
aorta (to body)
RIGHT SIDE
LEFT SIDE
atrium
atrium
pulmonary vein (from lungs)
valves (stop blood flowing backwards)
ventricle
ventricle

'Pacemaker' cells in the right atrium wall control resting heart rate.

An artificial pacemaker (electrical device put under the skin) can control heartbeat if the natural pacemaker cells don't work properly.

Coronary arteries supply the heart muscle with oxygenated blood.

Blood Vessels and Blood

Three Types of Blood Vessel

1. **Arteries carry blood away from the heart.**

lumen

thick muscle and elastic layers because blood pressure is high

2. **Capillaries carry blood close to body cells to exchange substances.**

thin, permeable walls to allow substances to diffuse in and out easily

3. **Veins carry blood back to the heart.**

valves inside stop blood flowing backwards

thinner walls than arteries because blood pressure is lower

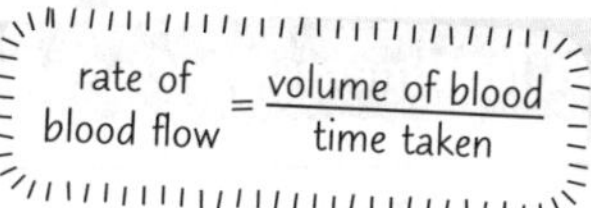
$\text{rate of blood flow} = \frac{\text{volume of blood}}{\text{time taken}}$

Four Blood Components

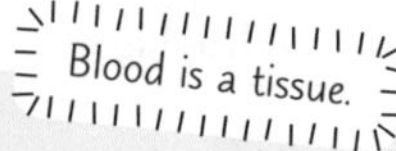
Blood is a tissue.

Component	Function	Description
1 Red blood cells	**Carry oxygen around the body.**	no nucleus so more room for O_2; biconcave shape = big surface area = lots of O_2 absorbed; contains haemoglobin, which binds to O_2
2 White blood cells	**Defend against infection.**	nucleus; → phagocytosis; → antitoxins; → antibodies
3 Platelets	**Help blood to clot at a wound.**	fragments of cells; a plate; a platelet
4 Plasma	**Carries everything in the blood.**	liquid; hormones, amino acids, red blood cells, urea, glucose, antibodies, antitoxins, white blood cells, proteins, platelets, CO_2

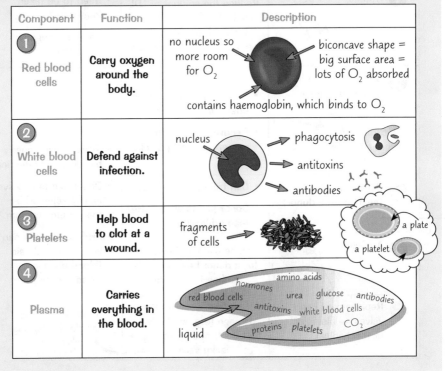

Cardiovascular Disease

Coronary Heart Disease

CARDIOVASCULAR DISEASES — diseases of the heart or blood vessels, e.g. coronary heart disease:

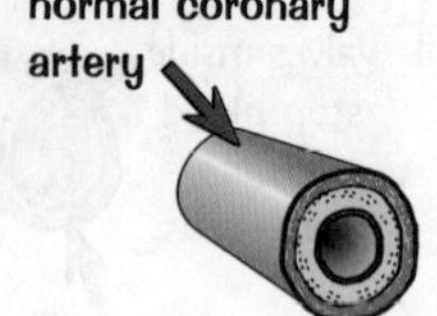
normal coronary artery

coronary artery of someone with coronary heart disease
fatty deposit
blood flow is restricted so there's a lack of oxygen to the heart muscle
This can cause a heart attack.

Five Treatments for Cardiovascular Disease

Treatment	Advantages	Disadvantages
1 Statins	Reduce the amount of LDL cholesterol in the blood, which slows down the formation of fatty deposits.	• Need to be taken long-term. • Can have negative side effects.
2 Stent (tube put in coronary artery) stent fatty deposit squashed	• Keeps coronary arteries open for a long time. • Recovery time from surgery is pretty quick.	• Surgery can cause bleeding and infection. • Donor hearts or valves can be rejected by the immune system. • Artificial devices can lead to thrombosis (blood clots in blood vessels).
3 Heart transplant (heart from a donor)	• Can treat heart failure. • Donor hearts work better than artificial ones.	
4 Artificial heart	Can be used while waiting for a donor heart or while the heart is healing.	
5 Replacement heart valves — biological or mechanical	Can treat severe valve damage (e.g. stiff valves that don't open properly or leaky valves).	

Faulty heart valves stop blood circulating effectively.

Health and Disease

Health

HEALTH — the state of physical and mental wellbeing.

Diseases can cause ill health.

These three things can also affect health:

1. **diet**  
2. **stress**  
3. **life situation, e.g. access to health care** 
HOSPITAL

Two Types of Disease

1. **COMMUNICABLE DISEASE — a disease that can spread from person to person or between animals and people.**

2. **NON-COMMUNICABLE DISEASE — a disease that cannot spread between people or between animals and people.**

Four Ways That Diseases May Interact

	Initial problem	Issue that can be made more likely
1	disorder affecting immune system	communicable diseases
2	infection by certain viruses	certain cancers
3	pathogen infection that causes an immune system reaction	allergic reactions (e.g. rashes or asthma)
4	severe physical health problems	mental health issues (e.g. depression)

Cost of Non-Communicable Diseases

Human cost —

Tens of millions of people die from non-communicable diseases each year.

Those living with these diseases may have a poorer quality of life and a shorter life-span.

Financial cost —

Researching and treating diseases costs health organisations (e.g. the NHS) loads of money.

Those with diseases may not be able to work, which can affect family finances as well as the country's economy.

Risk Factors for Diseases and Cancer

Five Risk Factors for Non-Communicable Diseases

RISK FACTORS — things that are linked to an increase in the likelihood that a person will develop a certain disease during their lifetime.

1. A lack of exercise and an unhealthy (e.g. high fat) diet are linked to cardiovascular disease.

2. Obesity is linked to Type 2 diabetes and cancer of the bowel, liver and kidneys.

3. Drinking too much alcohol can cause liver disease and affect brain function.

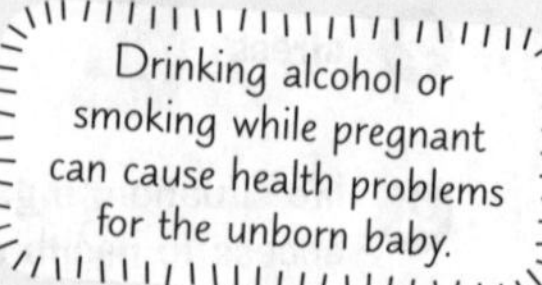
Drinking alcohol or smoking while pregnant can cause health problems for the unborn baby.

4. Smoking can cause cardiovascular disease, lung disease and lung cancer. It is also linked to mouth, bowel, stomach and cervical cancer.

5. Exposure to carcinogens (e.g. ionising radiation) can cause cancer.

Many non-communicable diseases are caused by several risk factors interacting.

Cancer

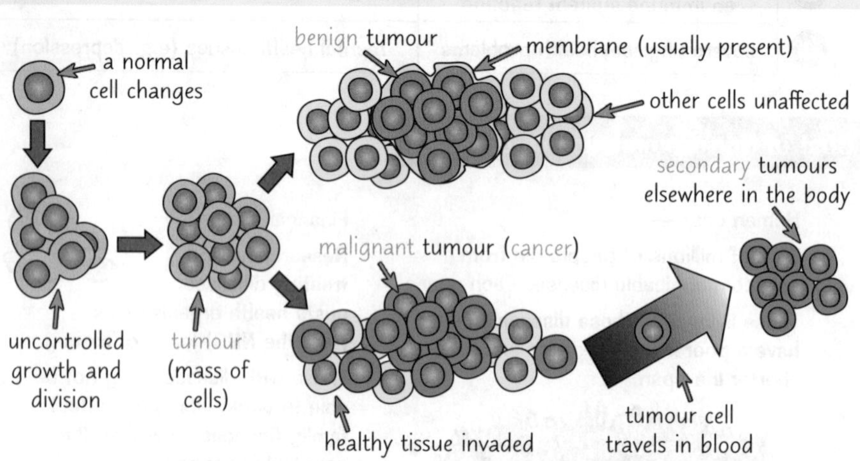
a normal cell changes
uncontrolled growth and division
tumour (mass of cells)
benign tumour
membrane (usually present)
other cells unaffected
malignant tumour (cancer)
healthy tissue invaded
tumour cell travels in blood
secondary tumours elsewhere in the body

Risk factors for cancer include lifestyle factors (e.g. smoking) as well as genetic factors — some people inherit faulty genes that make them more susceptible to cancer.

Plant Cell Organisation

Six Plant Tissues

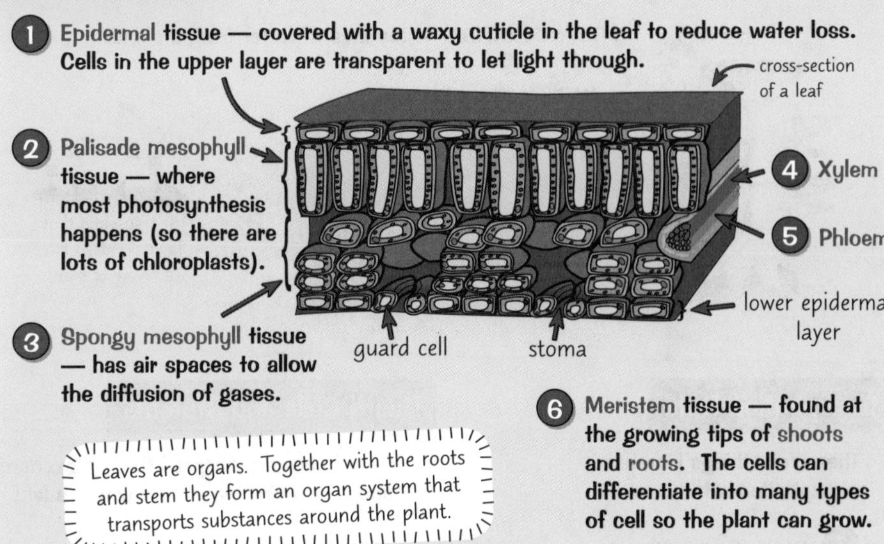
cross-section of a leaf
4 Xylem
5 Phloem
guard cell
stoma
lower epidermal layer

1. **Epidermal tissue — covered with a waxy cuticle in the leaf to reduce water loss. Cells in the upper layer are transparent to let light through.**
2. **Palisade mesophyll tissue — where most photosynthesis happens (so there are lots of chloroplasts).**
3. **Spongy mesophyll tissue — has air spaces to allow the diffusion of gases.**
4. Xylem
5. Phloem
6. **Meristem tissue — found at the growing tips of shoots and roots. The cells can differentiate into many types of cell so the plant can grow.**

Leaves are organs. Together with the roots and stem they form an organ system that transports substances around the plant.

Xylem

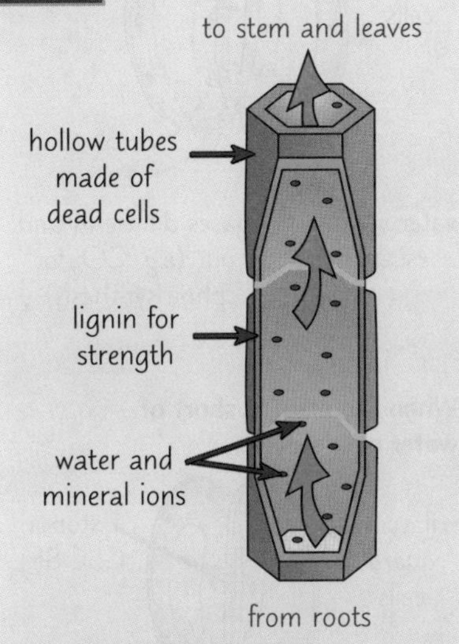
to stem and leaves
hollow tubes made of dead cells
lignin for strength
water and mineral ions
from roots

Xylem tissue carries water in the transpiration stream.

Phloem

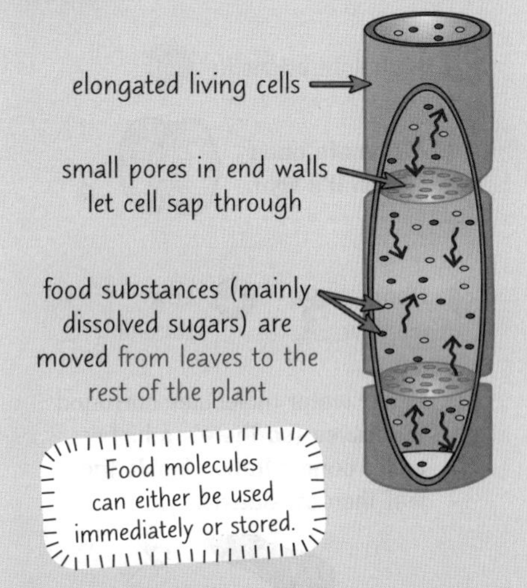
elongated living cells
small pores in end walls let cell sap through
food substances (mainly dissolved sugars) are moved from leaves to the rest of the plant
Food molecules can either be used immediately or stored.

TRANSLOCATION — the process in which food is moved through phloem tubes.

Transpiration

Transpiration Basics

TRANSPIRATION — the loss of water from a plant.

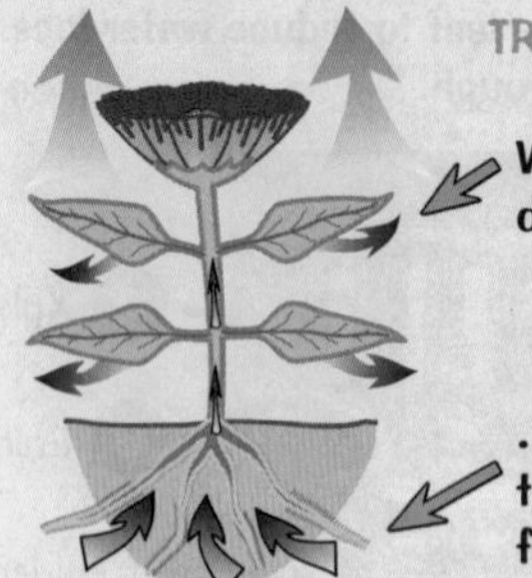

Water evaporates and diffuses out of the plant...

...which causes more water to be drawn into the plant from the roots.

It transpired that Jay did indeed have a favourite plant.

Transpiration Rate

These four things increase transpiration rate:

Warm temperatures

Water molecules have more energy.

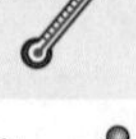

High light intensity

Stomata open when it's light.

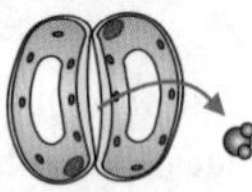

Good air flow

Low humidity

Fewer water molecules surround the leaves (so there's a higher water concentration inside the leaf than outside it).

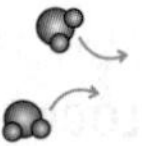

Guard Cells and Stomata

Guard cells are adapted for gas exchange and controlling water loss within a leaf.

When the plant has lots of water...

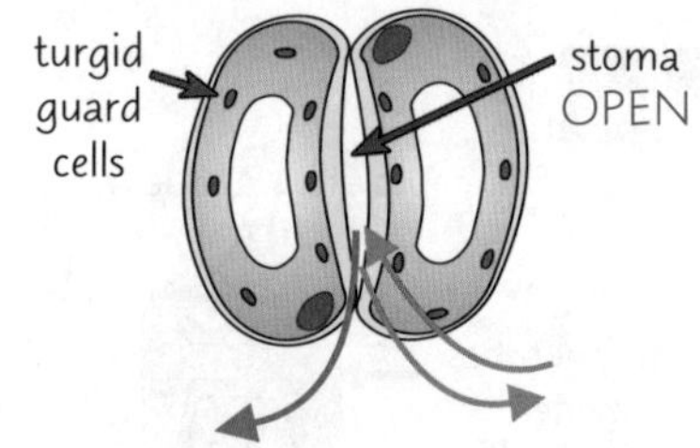

turgid guard cells
stoma OPEN
water vapour escapes
gases diffuse in and out (e.g. CO_2 for photosynthesis)

When the plant is short of water or it's dark...

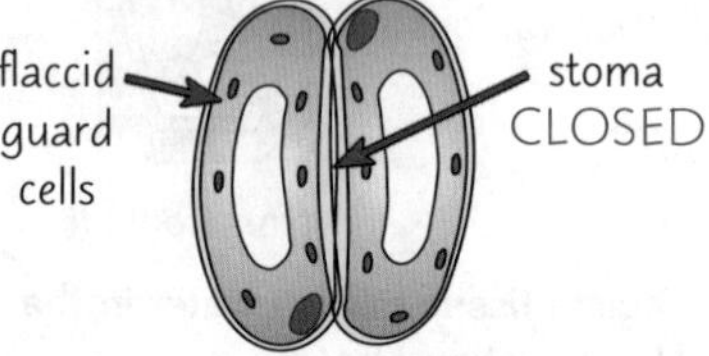

flaccid guard cells
stoma CLOSED

Communicable Disease

Infectious Diseases

PATHOGENS — microorganisms that cause diseases which spread between organisms.

Disease	Pathogen	How it's spread	Symptoms	Prevention / treatment
Rose black spot	Fungus	• Water • Wind	• Purple/black spots on leaves, can turn yellow and drop off • Reduced growth	• Removing and destroying infected leaves • Fungicides
Malaria	Protist	Mosquito vectors	• Fever • Can be fatal	• Mosquito nets • Stop mosquitoes from breeding
Salmonella food poisoning	Bacterium	Eating contaminated food	• Fever • Stomach cramps • Vomiting • Diarrhoea	• Vaccination of poultry • Hygienic food preparation
Gonorrhoea	Bacterium	Sexual contact	• Pain when urinating • Yellow/green discharge from vagina or penis	• Condoms • Antibiotics
Measles	Virus	Airborne droplets (coughs and sneezes)	• Fever • Red skin rash • Can be fatal	Vaccination of children
HIV	Virus	• Sexual contact • Exchanging bodily fluids (e.g. blood)	• Flu-like (initially) • A damaged immune system (late stage infection/AIDS)	• Condoms • Avoid sharing needles • Antiretrovirals
Tobacco mosaic virus (TMV)	Virus	It can be spread by direct contact between plants but you don't need to learn this for TMV.	Mosaic pattern on leaves, which reduces photosynthesis and growth	You don't need to know how TMV is prevented. Have they tried destroying all the world's tiles?

Fighting Disease

Four Non-Specific Defence Systems Against Pathogens

1. **Skin** — acts as a barrier and secretes antimicrobial substances to kill pathogens.
2. **Nose** — hairs and mucus trap particles containing pathogens.
3. **Trachea and bronchi** — mucus traps pathogens, and cilia waft mucus up to the throat so that it can be swallowed.
4. **Stomach** — hydrochloric acid kills pathogens.

Three Ways White Blood Cells Attack Pathogens

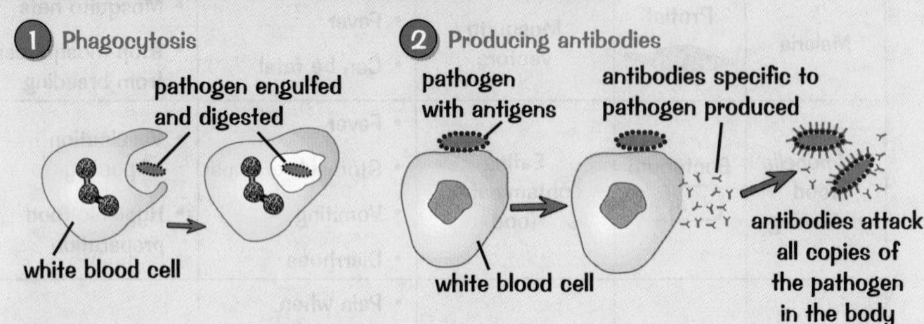

1 Phagocytosis
pathogen engulfed and digested
white blood cell
2 Producing antibodies
pathogen with antigens
antibodies specific to pathogen produced
white blood cell
antibodies attack all copies of the pathogen in the body

3. **Producing antitoxins** — these counteract toxins produced by invading bacteria.

Vaccination

Vaccinating a large proportion of the population greatly reduces the spread of pathogens so that even people who aren't vaccinated are unlikely to catch the disease.

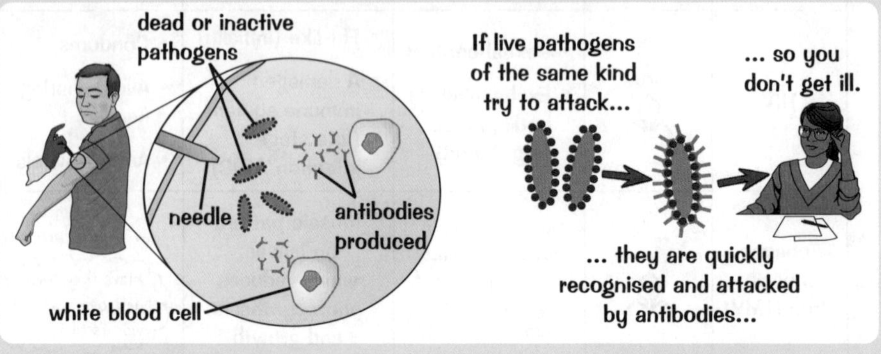

dead or inactive pathogens
needle
antibodies produced
white blood cell
If live pathogens of the same kind try to attack...
... they are quickly recognised and attacked by antibodies...
... so you don't get ill.

Drugs

Types of Drugs

Antibiotics kill bacteria.

Specific antibiotics kill specific types of bacteria.

Use of antibiotics
Deaths from bacterial disease
Time

It's hard to develop drugs that destroy viruses because they live and reproduce inside cells.

Painkillers treat the symptoms of disease but don't kill pathogens.

Where Drugs Come From

Drug	Type	Source
Digitalis	Heart drug	Foxgloves
Aspirin	Painkiller	Willow
Penicillin	Antibiotic	*Penicillium* mould

Nowadays, drugs are often synthesised in a lab, but the process still might start with a chemical extracted from a plant.

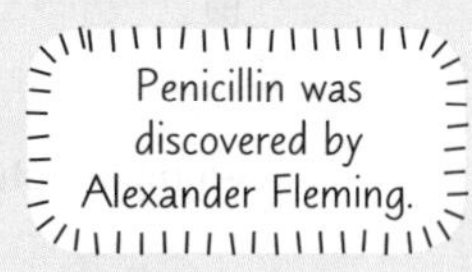
Penicillin was discovered by Alexander Fleming.

Drug Testing

New drugs are trialled to check they're safe and effective.
They are tested for three things:

1. Toxicity — how harmful the drug is.
2. Efficacy — whether the drug works and produces the effect you're looking for.
3. Dosage — the concentration that should be given, and how often it should be given.

Preclinical Testing

Tests on human cells and tissues → Tests on live animals

Clinical Trials

Tests on healthy volunteers

The dosage is gradually increased from a very low initial dose

↓

Tests on ill patients

Finding optimum dose

↓

Peer review

Clinical trials are often double-blind

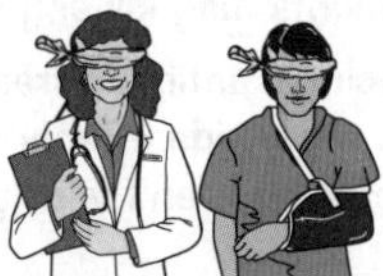

Given drug

Given placebo

PLACEBOS — substances that are like the drug being tested but don't do anything.

Monoclonal Antibodies

Monoclonal Antibodies

Monoclonal antibodies are produced from lots of clones of a single white blood cell (a B-lymphocyte).

Monoclonal antibodies

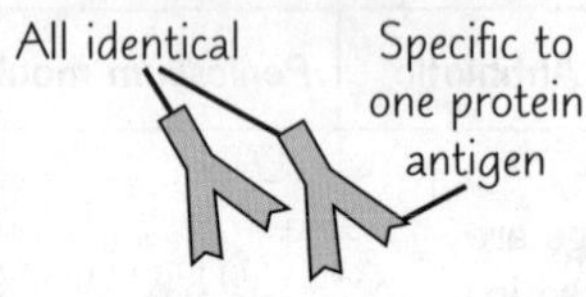
All identical
Specific to one protein antigen

B-lymphocyte

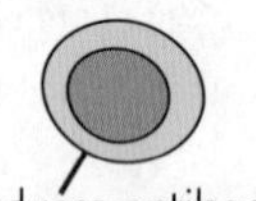

Produces antibodies

Producing Monoclonal Antibodies

Mouse injected with chosen antigen

B-lymphocytes from mouse — make antibodies but don't divide easily

Tumour cells in lab — don't make antibodies but divide lots

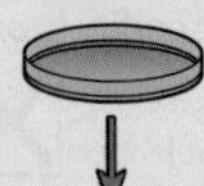

B-lymphocyte fused with tumour cell

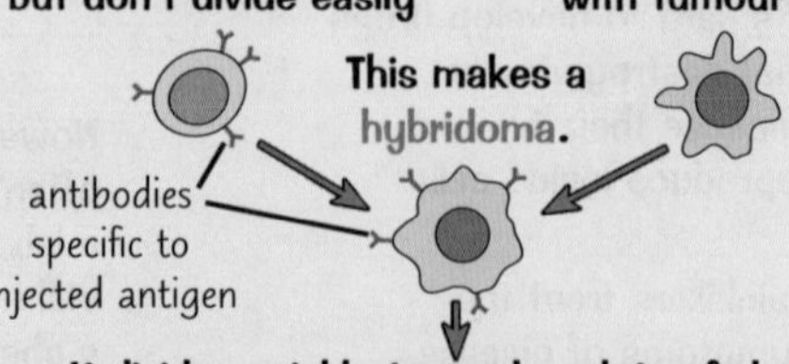
antibodies specific to injected antigen

This makes a hybridoma.

It divides quickly to produce lots of clones that produce the monoclonal antibodies.

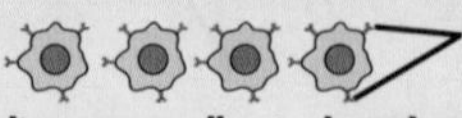
monoclonal antibodies

The antibodies are collected and purified.

Three Uses of Monoclonal Antibodies

1 Cancer treatment

monoclonal antibody specific to tumour marker

attached

toxic drug/radioactive substance/ chemical to stop cells growing and dividing

antibody binds to tumour markers

substance is delivered to cancer cell

antigen

cancer cell

normal body cells are unharmed

tumour marker unique to cancer cells

2 Locating specific molecules in research

fluorescent dye
antibody binds to antigen on target molecule and dye is detected
antigen

+ : Monoclonal antibodies target cancer cells, whereas other treatments may kill any cells.

– : Monoclonal antibody treatments cause more side effects than expected so aren't widely used.

3 Measuring levels of substances in blood or urine

E.g. pathogens, hormones and other chemicals.

Pregnancy tests work by detecting hormones in urine.

Plant Diseases and Defences

Six Signs of Plant Disease

1. Stunted growth
2. Spots on leaves
3. Discolouration
4. Patches of decay (rot)
5. Malformed stems or leaves
6. Abnormal growths, e.g. lumps

Seeing pests on a plant is a sign of an infestation.

Causes of Plant Diseases

Plants can be:

Infected by pathogens		
Viruses	Bacteria	Fungi

Infested by
Insects (e.g. aphids)

Abducted
by aliens...

Affected by
Nitrate deficiency, which stunts growth — nitrate ions are needed to make proteins for growth
Magnesium deficiency, which causes chlorosis (yellow leaves) — magnesium ions are needed to make chlorophyll

Three Ways To Identify Plant Disease

1. Look up the signs in a gardening manual or on a gardening website.

2. Take the infected plant to a lab, where scientists can identify the pathogen.

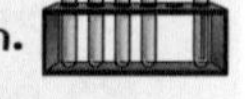

3. Use testing kits that identify the pathogen using monoclonal antibodies.

Plant Defences

Physical	Mechanical	Chemical
Waxy cuticle on leaves — barrier to pathogens.	Thorns and hairs stop animals touching and eating plants.	Antibacterial chemicals kill bacteria and prevent disease.
Layers of dead cells around stems (e.g. bark) — barrier to pathogens.	Leaves that droop or curl when touched can knock insects off.	Some plants produce poisons to deter herbivores.
Cellulose cell walls — barrier to pathogens around cells.	Plants mimic other organisms to trick animals into not eating them.	

Photosynthesis

The Photosynthesis Reaction

PHOTOSYNTHESIS — an endothermic reaction in which energy is transferred to chloroplasts from the environment by light.

$$\text{carbon dioxide} + \text{water} \xrightarrow{\text{light}} \text{glucose} + \text{oxygen}$$

$$6CO_2 \quad 6H_2O \quad C_6H_{12}O_6 \quad 6O_2$$

Four Uses of Glucose in Plants

1. **Respiration — energy is transferred from glucose.**
2. **Strengthening cell walls — glucose is converted into cellulose, which is used to make strong cell walls.**
3. **Protein synthesis — glucose and nitrate ions (from soil) are used to make amino acids, which are then made into proteins.**
4. **Energy storage — glucose is turned into lipids or insoluble starch to store energy.**

Rate of Photosynthesis

An increase in any of these four factors tends to increase the rate of photosynthesis:

1. **Light intensity**
2. **Temperature**
3. **CO_2 concentration**
4. **Amount of chlorophyll**

Any of these factors can become the limiting factor (thing that stops photosynthesis happening any faster).

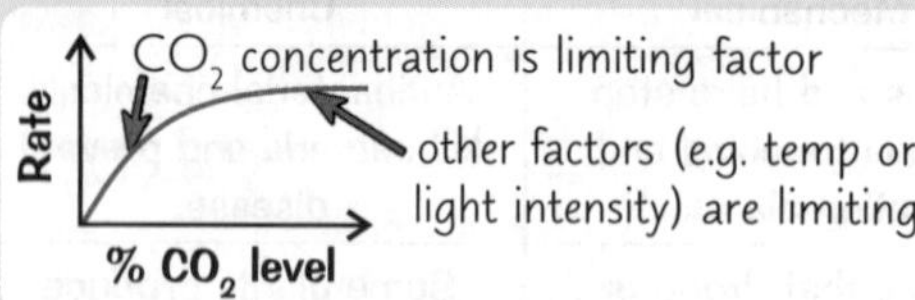
CO_2 concentration is limiting factor
other factors (e.g. temp or light intensity) are limiting
Rate
% CO_2 level

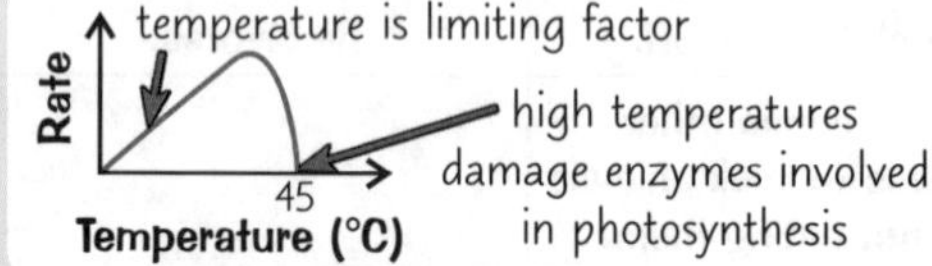
temperature is limiting factor
high temperatures damage enzymes involved in photosynthesis
Rate
45
Temperature (°C)

Inverse Square Law

The inverse square law links light intensity with distance from a light source:

$$\text{Light intensity} \propto \frac{1}{\text{distance (d)}^2}$$

E.g. doubling the distance means light intensity is four times smaller.

Greenhouses

You can control limiting factors in a greenhouse to maximise the rate of photosynthesis, but this costs money.

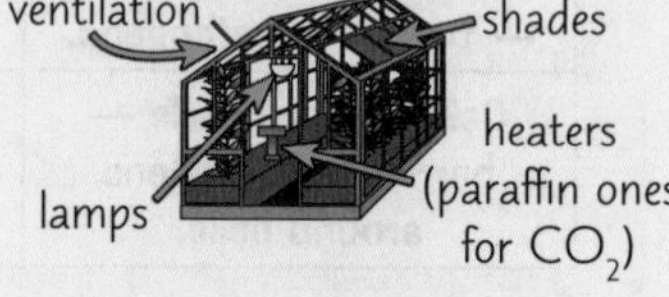
ventilation
shades
heaters (paraffin ones for CO_2)
lamps

Respiration

Energy Transfer

RESPIRATION — the process of transferring energy from glucose.
It's an exothermic reaction that goes on continuously in living cells.

The energy transferred is used for all living processes.
Three examples of these are:

1. To contract muscles for movement.

2. To keep warm (in mammals and birds).

3. To build up larger molecules from smaller ones.

Aerobic Respiration

AEROBIC RESPIRATION — respiration using oxygen.
It's the most efficient type of respiration.

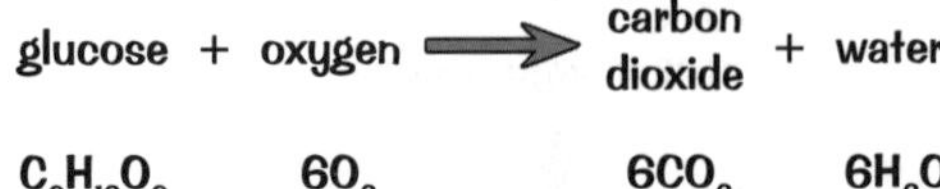

glucose + oxygen → carbon dioxide + water

$C_6H_{12}O_6$ + $6O_2$ → $6CO_2$ + $6H_2O$

Anaerobic Respiration

ANAEROBIC RESPIRATION — respiration without oxygen.
It transfers much less energy than aerobic respiration because glucose isn't fully oxidised.

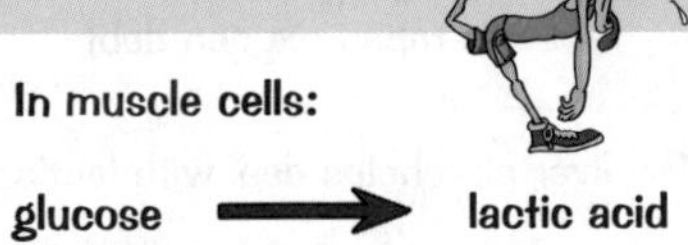

In muscle cells:

glucose → lactic acid

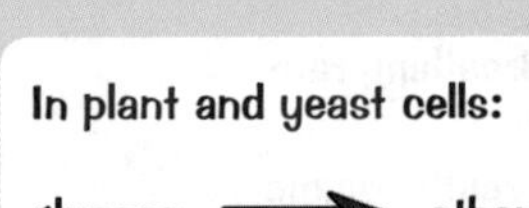

In plant and yeast cells:

glucose → ethanol + carbon dioxide

In yeast cells, anaerobic respiration is called fermentation.
The process is used to make bread and alcoholic drinks.

Metabolism and Exercise

Metabolism

METABOLISM — the sum of all the reactions that happen in a single cell or the body.

Metabolic reactions use energy from respiration to make new molecules, e.g.:

Glucose and nitrate ions combined to make amino acids, then made into proteins.

Lipids broken down into glycerol and fatty acids.

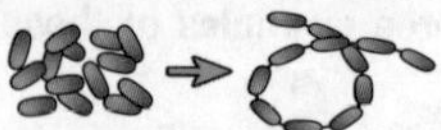

Glucose molecules joined together to make bigger carbohydrates (e.g. starch or glycogen).

Effects of Exercise on the Body

EXERCISE

more energy needed

more aerobic respiration needed

more oxygen needed

These three things increase to get more oxygen to your muscles:

1

Heart rate

2

Breathing rate

3

Breath volume

Long periods of exercise also cause muscle fatigue — the muscles get tired and stop contracting efficiently.

Oxygen Debt

OXYGEN DEBT — the amount of extra oxygen needed to react with built-up lactic acid and remove it from cells.

During vigorous exercise, not enough oxygen is supplied to the muscles so:

Anaerobic respiration takes place in muscles

Lactic acid builds up in muscles

Oxygen debt created

Heart rate and breathing rate stay high after exercise to repay oxygen debt

The liver also helps deal with lactic acid:

blood → muscle → blood with lactic acid → liver → lactic acid converted to glucose

Homeostasis and the Nervous System

Maintaining a Stable Internal Environment

HOMEOSTASIS — the regulation of the conditions inside your body and cells. It maintains a stable internal environment in response to changes in internal and external conditions.

Homeostasis maintains optimal conditions for enzyme action.

Component of control systems	Function of component
Receptors	To detect stimuli (changes in environment)
Coordination centres	To receive and process information from receptors and organise a response
Effectors	To produce a response to counteract change and restore optimal conditions

Automatic control systems can involve nervous or chemical responses.
Three things in your body that are maintained by control systems:

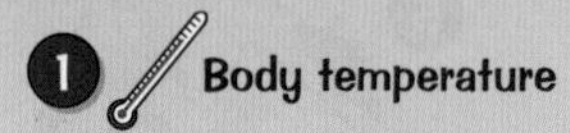
1
Body temperature

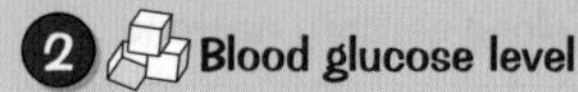
2
Blood glucose level

3
Water content

The Nervous System

NEURONES — cells that carry information as electrical impulses in the nervous system. The nervous system means that humans can react to their surroundings and coordinate their behaviour.

All this revision's really getting on my neurones...

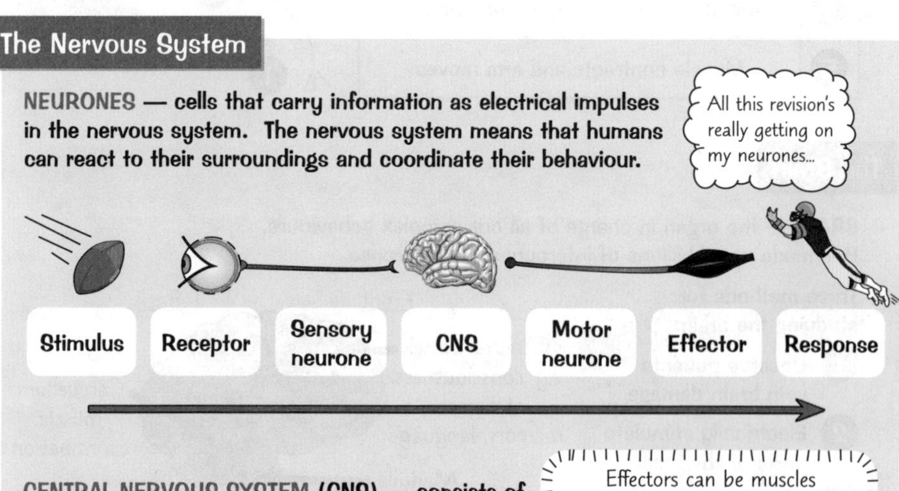
Stimulus
Receptor
Sensory neurone
CNS
Motor neurone
Effector
Response

CENTRAL NERVOUS SYSTEM (CNS) — consists of the brain and spinal cord. It is connected to the body by sensory neurones and motor neurones.

Effectors can be muscles (which respond to nervous impulses by contracting) or glands (which secrete hormones).

Synapses, Reflexes and the Brain

Synapses

SYNAPSE — the connection between two neurones.
A nerve signal is transferred across a synapse by the diffusion of chemicals.

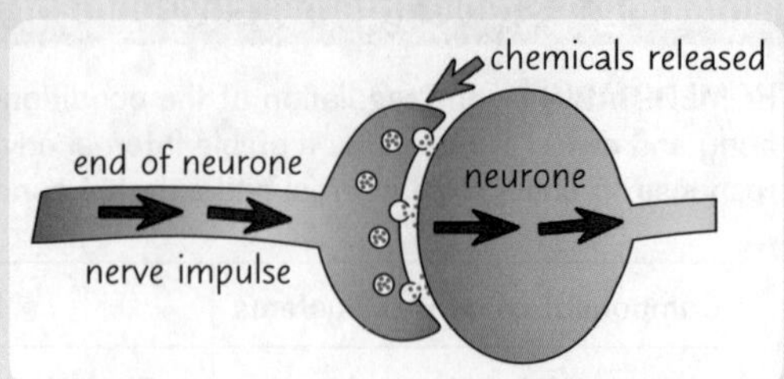

chemicals released
end of neurone
neurone
nerve impulse

Reflex Arcs

REFLEXES — rapid, automatic responses to certain stimuli that don't involve the conscious part of the brain. They can reduce the chance of injury.

Five steps in a reflex arc:

1	**Stimulation of pain receptor**
2	**Impulses travel along sensory neurone**
3	**Impulses passed along relay neurone**
4	**Impulses travel along motor neurone**
5	**Muscle contracts and arm moves**

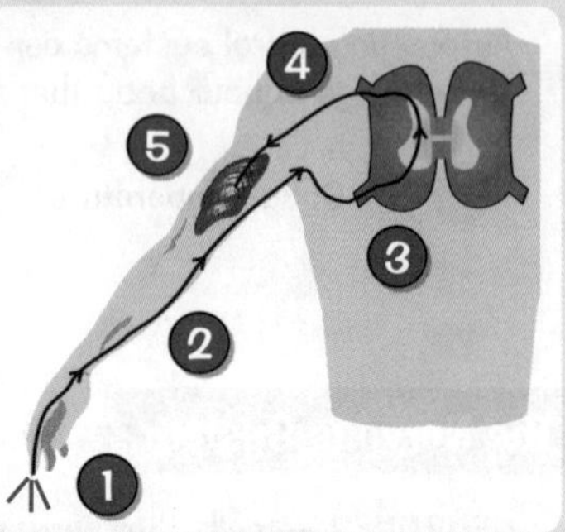

4
5
3
2
1

The Brain

BRAIN — the organ in charge of all our complex behaviours.
It is made up of billions of interconnected neurones.

Three methods for studying the brain:

1. **Observe patients with brain damage**
2. **Electrically stimulate parts of the brain**
3. **Use MRI scanners**

Front
Back
Cerebral cortex e.g. consciousness, intelligence, memory, language
Cerebellum muscle coordination
Medulla unconscious activities
Spinal cord

The brain is complex and delicate, so investigating or treating it is difficult and risky.

The Eye

Structure of the Eye

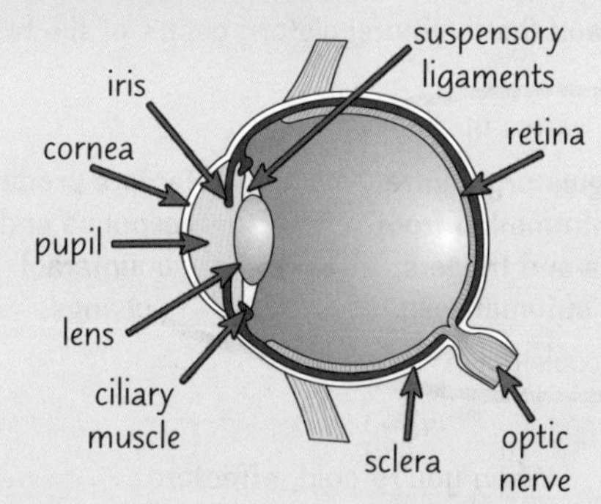
suspensory
ligaments
iris
retina
cornea
pupil
lens
ciliary
muscle
sclera
optic
nerve

Sclera	Tough, supporting wall
Cornea	Transparent outer layer at front
Iris	Contains muscles controlling pupil size
Retina	Contains receptor cells sensitive to light intensity and colour
Optic nerve	Carries impulses from receptor cells to the brain

Eye Reflexes

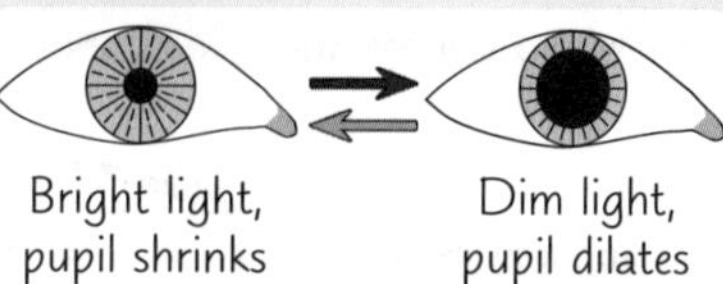

Bright light, pupil shrinks

Dim light, pupil dilates

ACCOMMODATION — the lens changing shape to focus light on the retina.

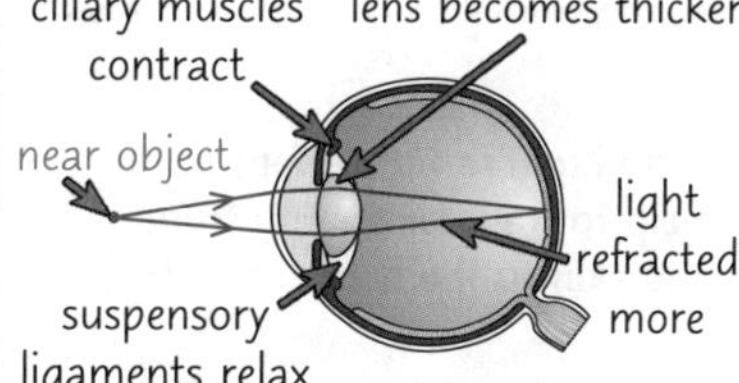
ciliary muscles
contract
lens becomes thicker
near object
light
refracted
more
suspensory
ligaments relax

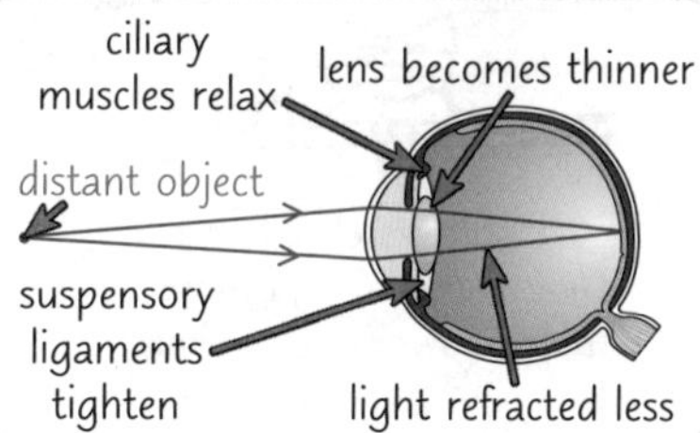
ciliary
muscles relax
lens becomes thinner
distant object
suspensory
ligaments
tighten
light refracted less

Common Defects of the Eye

Long-sightedness (hyperopia):

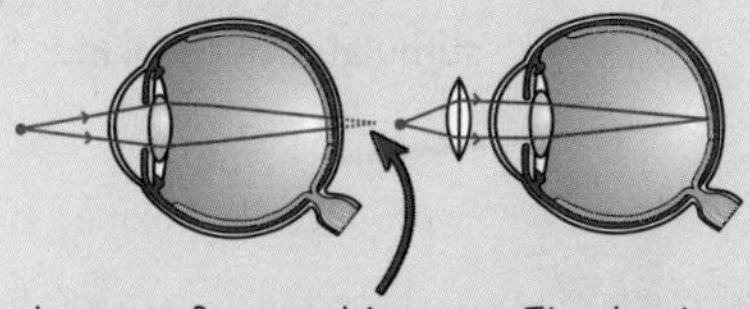

Image of near objects brought into focus behind retina.

Fixed using convex lens.

Short-sightedness (myopia):

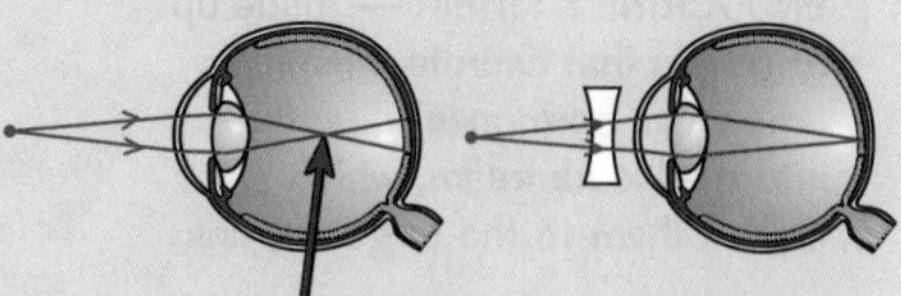

Image of distant objects brought into focus in front of retina.

Fixed using concave lens.

These defects are usually treated by wearing glasses — three other options are:

1.  **contact lenses**
2.  **laser eye surgery**
3.  **replacement lens surgery**

Body Temperature and Hormones

Controlling Body Temperature

There are temperature receptors in the skin and the thermoregulatory centre of the brain.

Temperature receptors detect core body temperature is too high or too low. → **Thermoregulatory centre receives information from receptors and triggers effectors automatically.** → **Effectors produce response and counteract change.**

body warms up

body cools down

When you're hot, effectors help to transfer energy from the skin to the environment:

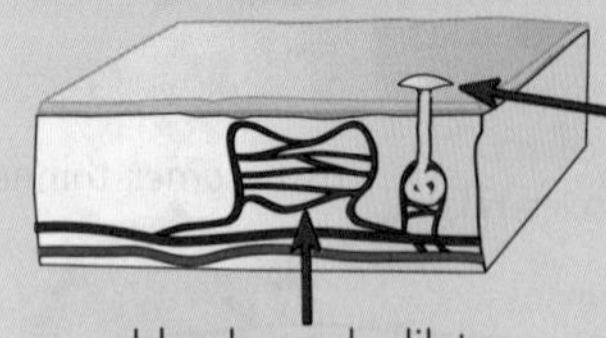

sweat gland produces sweat

blood vessels dilate (vasodilation) and more blood flows close to skin

When you're cold, effectors reduce the energy transferred to the environment:

no sweat

blood vessels constrict (vasoconstriction) and less blood flows close to skin

You also shiver when you're cold — this is your skeletal muscles contracting.

The Endocrine System

ENDOCRINE SYSTEM — made up of glands that secrete chemicals (known as hormones) directly into the bloodstream, which carries them to the target organs.

Pituitary gland 'master gland', stimulates other glands

Thyroid produces thyroxine

Adrenal gland produces adrenaline

Pancreas produces insulin

Ovaries (female) produce oestrogen

Testes (male) produce testosterone

The effects of hormones are slower than nerves but last longer.

Blood Glucose and Hormones

Controlling Blood Glucose

Five steps to reduce blood glucose:

1. Blood with too much glucose
2. Pancreas detects high blood glucose and secretes insulin
3. Insulin causes glucose to move into cells
4. Insulin makes the liver turn glucose into glycogen, which is stored in the liver and muscles
5. Blood glucose reduced

Five steps to increase blood glucose:

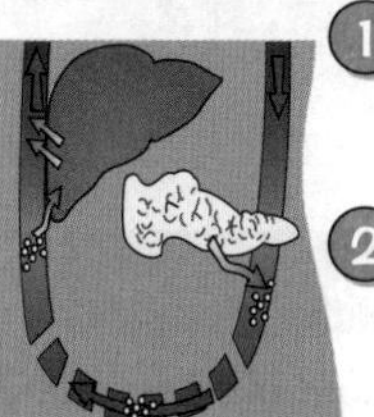

1. Blood with too little glucose
2. Glucagon secreted by the pancreas
3. Too little glucose but glucagon as well
4. Glucagon makes the liver turn glycogen into glucose, which is released from the liver
5. Blood glucose increased

Insulin and glucagon work in a negative feedback cycle.

Diabetes

	Type 1	Type 2
Cause	Pancreas produces little or no insulin	Cells no longer respond to insulin properly
Effect	Blood glucose can rise to dangerously high levels	
Treatment	Insulin injections	Carbohydrate-controlled diet and regular exercise

Obesity is a major risk factor for Type 2 diabetes.

Adrenaline and Thyroxine

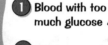

Adrenaline released in response to fear or stress → Increases heart rate → Increases supply of O_2 and glucose to muscles and brain → Readies body for 'fight or flight'

Thyroxine plays a role in regulating the basal metabolic rate and is important for protein synthesis for growth and development.

Thyroxine levels are controlled by negative feedback.

Waste Substances and the Kidneys

Waste Substances

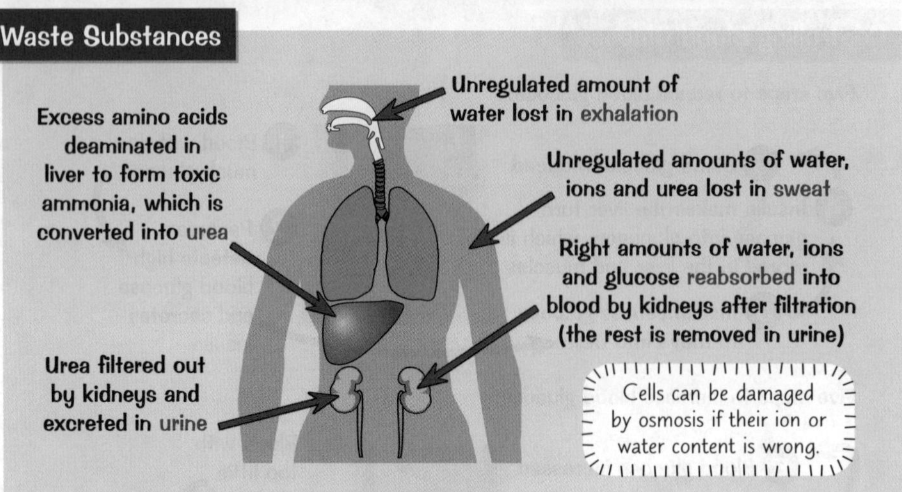
Unregulated amount of water lost in exhalation
Excess amino acids deaminated in liver to form toxic ammonia, which is converted into urea
Unregulated amounts of water, ions and urea lost in sweat
Right amounts of water, ions and glucose reabsorbed into blood by kidneys after filtration (the rest is removed in urine)
Urea filtered out by kidneys and excreted in urine
Cells can be damaged by osmosis if their ion or water content is wrong.

Controlling the Concentration of Urine

The concentration of urine is controlled by anti-diuretic hormone (ADH).

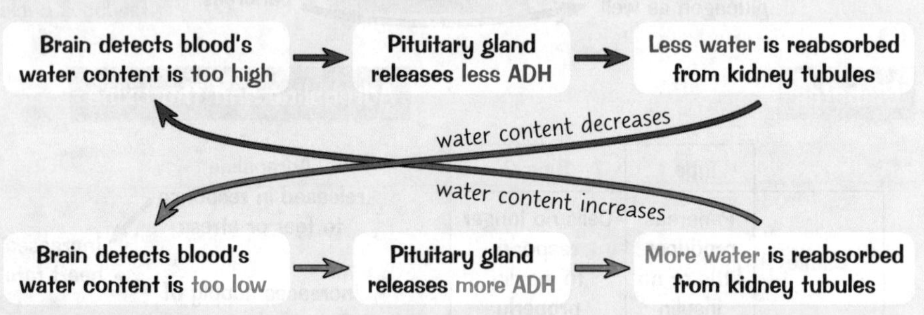
Brain detects blood's water content is too high
Pituitary gland releases less ADH
Less water is reabsorbed from kidney tubules
water content decreases
water content increases
Brain detects blood's water content is too low
Pituitary gland releases more ADH
More water is reabsorbed from kidney tubules

Kidney Failure

People with kidney failure need to have regular dialysis.

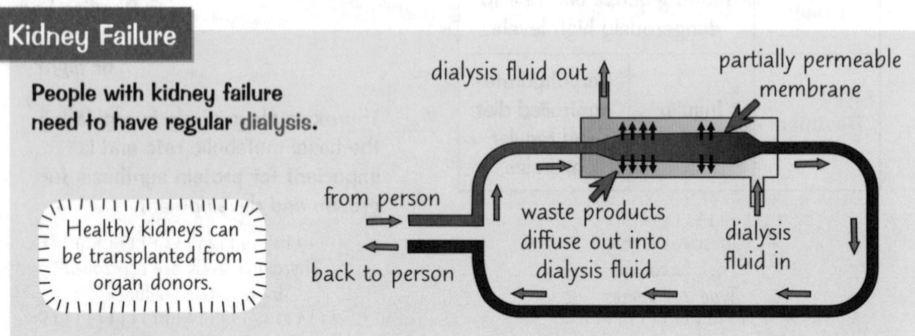
Healthy kidneys can be transplanted from organ donors.
dialysis fluid out
partially permeable membrane
from person
back to person
waste products diffuse out into dialysis fluid
dialysis fluid in

Puberty and the Menstrual Cycle

Puberty and Sex Hormones

PUBERTY — when the body starts releasing sex hormones, which trigger the development of secondary sexual characteristics (e.g. facial hair in men and breasts in women).

In men, the main reproductive hormone is testosterone, which stimulates sperm production.

In women, the main reproductive hormone is oestrogen.

The Menstrual Cycle

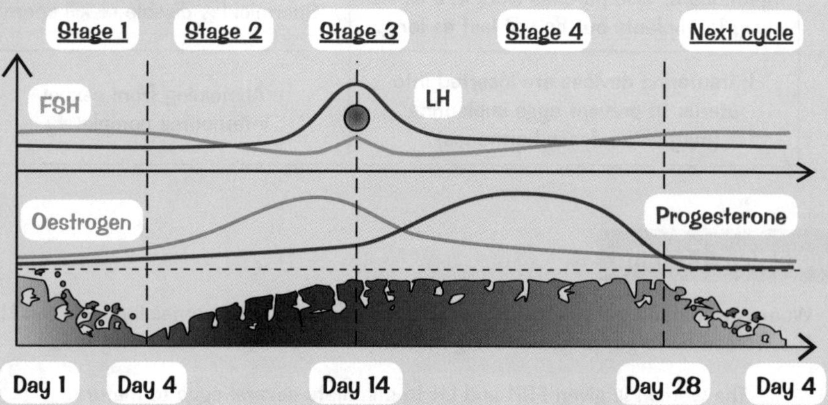
Stage 1
Stage 2
Stage 3
Stage 4
Next cycle
FSH
LH
Oestrogen
Progesterone
Day 1
Day 4
Day 14
Day 28
Day 4

Stage 1 — Menstruation starts. Uterus lining breaks down.

Stage 2 — Uterus lining builds up into thick spongy layer full of blood vessels ready to receive a fertilised egg.

Stage 3 — Egg develops and is released from ovary — this is called ovulation.

Stage 4 — Wall is maintained. If no fertilised egg lands on wall, lining breaks down and cycle starts again.

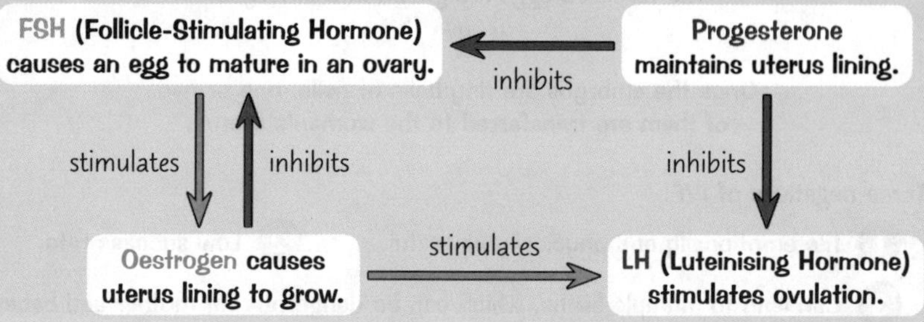
FSH (Follicle-Stimulating Hormone) causes an egg to mature in an ovary.
inhibits
Progesterone maintains uterus lining.
stimulates
inhibits
inhibits
Oestrogen causes uterus lining to grow.
stimulates
LH (Luteinising Hormone) stimulates ovulation.

Controlling Fertility

Reducing Fertility

CONTRACEPTION — methods of reducing the likelihood of sperm reaching an ovulated egg.

Hormonal methods	Non-hormonal methods
Oral contraceptive pills contain hormones that inhibit FSH and stop eggs maturing.	Condoms and diaphragms physically prevent sperm from reaching egg.
The contraceptive implant releases progesterone continuously to stop maturation and release of eggs.	Sterilisation — a permanent surgical procedure to stop a man or woman from being fertile.
Injections or skin patches work in a similar way to implants but do not last as long.	Spermicides disable or kill sperm.
Intrauterine devices are inserted into uterus to prevent eggs implanting (may also release hormones).	Abstaining from sexual intercourse completely.

Increasing Fertility

Women who can't get pregnant can be given a fertility drug containing FSH and LH. If a woman can't get pregnant using medication, she may choose to try IVF:

The woman is given FSH and LH to stimulate several eggs to mature.

The eggs are collected from the woman's ovaries.

IVF stands for "*in vitro* fertilisation".

The eggs are fertilised in a lab using the man's sperm.

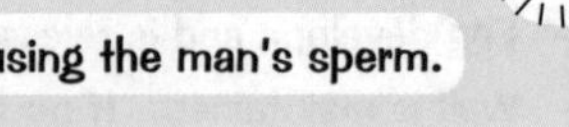

The fertilised eggs are grown into embryos.

Once the embryos are tiny balls of cells, one or two of them are transferred to the woman's uterus.

Three negatives of IVF:

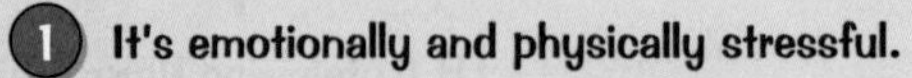

1. It's emotionally and physically stressful.

2. Low success rate.
3. Can lead to multiple births, which can be dangerous for mother and babies.

Plant Hormones

Phototropism

AUXIN — a plant hormone that controls growth near the tips of shoots and roots.

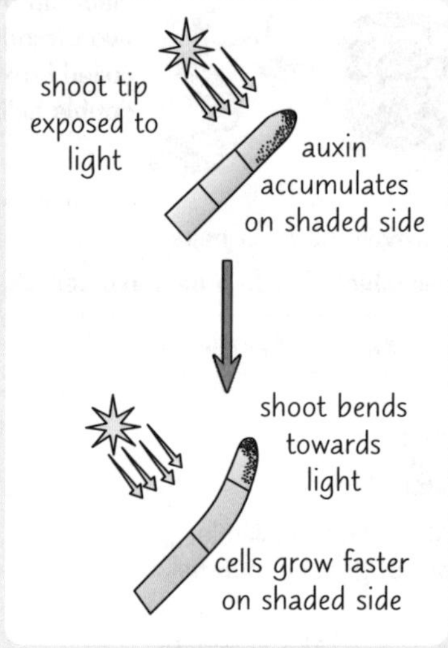
shoot tip exposed to light
auxin accumulates on shaded side
shoot bends towards light
cells grow faster on shaded side

Gravitropism/Geotropism

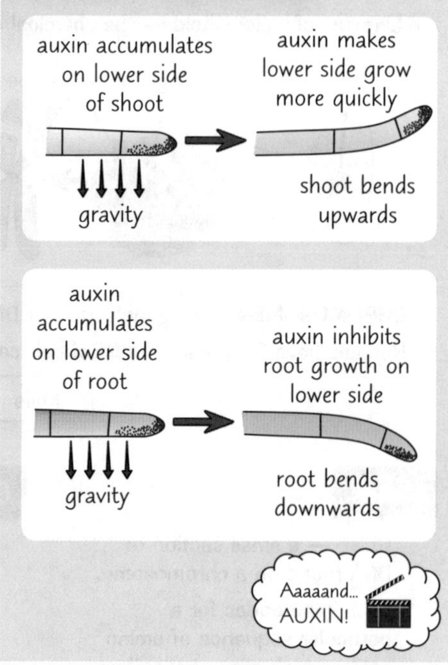
auxin accumulates on lower side of shoot
gravity
auxin makes lower side grow more quickly
shoot bends upwards
auxin accumulates on lower side of root
gravity
auxin inhibits root growth on lower side
root bends downwards
Aaaaand... AUXIN!

Commercial Uses of Plant Hormones

Auxins	Gibberellin	Ethene
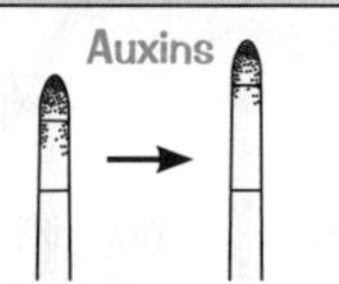	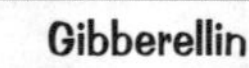 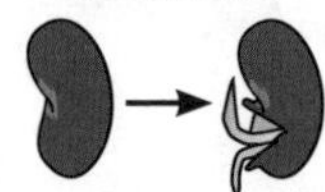	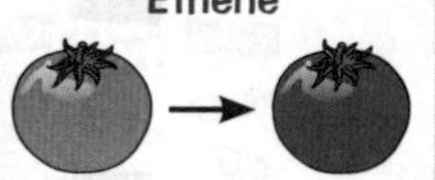
Developed to selectively kill weeds whilst leaving crops untouched	**Stimulates seeds to germinate at any time of year**	**Speeds up ripening of fruits in transport to shops**
Added to rooting powders to promote root growth in plant cuttings	**Induces flowering without the need for specific conditions**	**Its effects can be blocked to delay ripening in storage**
Stimulates cell division in tissue culture so plants can be cloned from a few cells	**Added to fruits to make them grow larger**	

Ethene also controls cell division.

DNA

Genetic Material

DeoxyriboNucleic Acid — the chemical a cell's nuclear genetic material is made from.

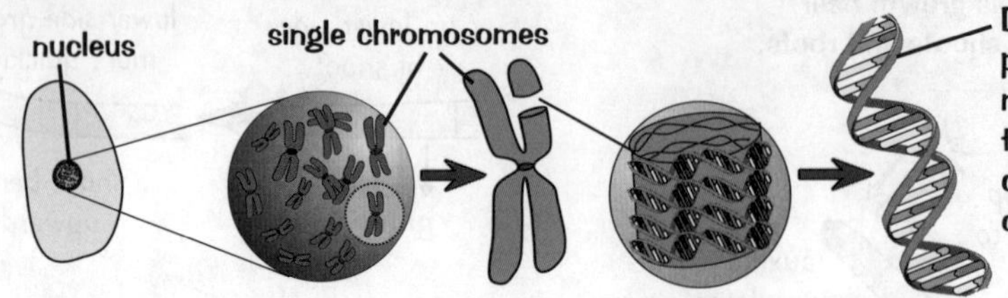
nucleus
single chromosomes

DNA is a polymer made up of two strands coiled into a double helix.

CHROMOSOMES — long molecules of DNA that normally come in pairs.

Humans have 23 pairs. The 23rd pair carries genes which decide a person's sex.

XY	Male

XX	Female

Genes

GENE — a small section of DNA found on a chromosome.

Each gene codes for a particular sequence of amino acids, which are put together to make a specific protein.

20 different amino acids

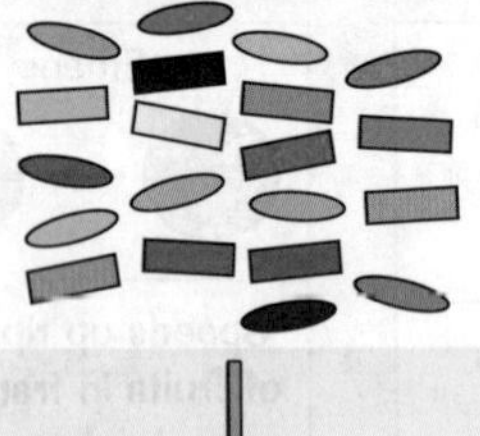

1000s of possible proteins

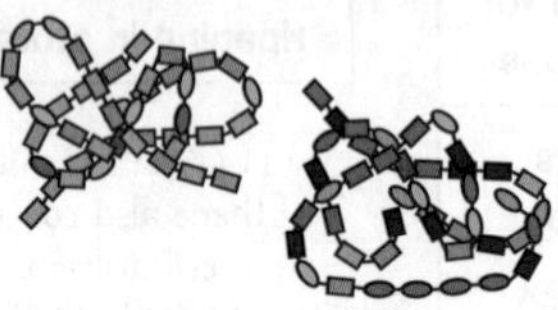

Genomes

GENOME — an organism's entire set of genetic material.

The complete human genome has been worked out.

Genes linked to diseases can be identified.

This helps us better understand inherited diseases, so we can develop effective treatments.

Tiny differences in people's genomes can be studied.

This helps us trace the migration patterns of past human populations.

DNA, Proteins and Mutations

The Structure of DNA

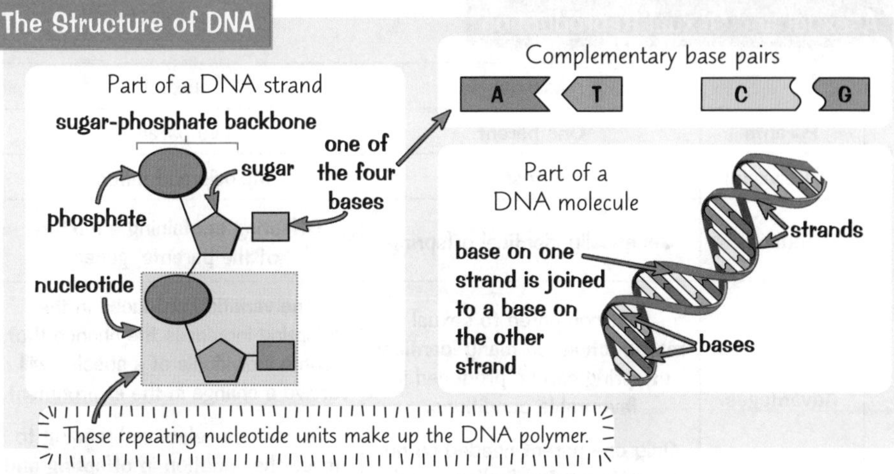
Part of a DNA strand
sugar-phosphate backbone
phosphate
sugar
one of the four bases
nucleotide
Complementary base pairs
A
T
C
G
Part of a DNA molecule
strands
base on one strand is joined to a base on the other strand
bases

These repeating nucleotide units make up the DNA polymer.

Protein Synthesis

Each amino acid is coded for by a sequence of three bases.

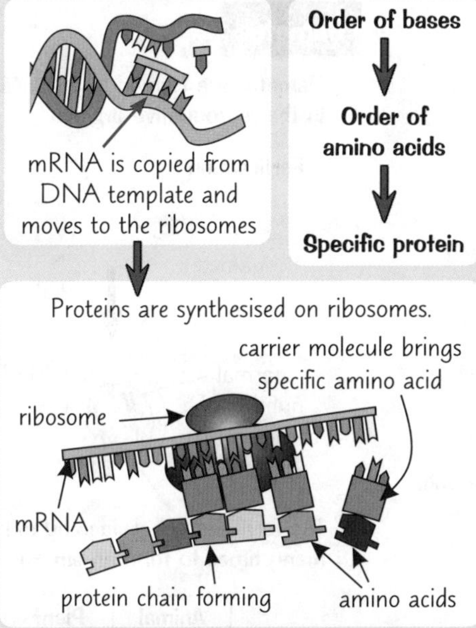
mRNA is copied from DNA template and moves to the ribosomes
Order of bases
Order of amino acids
Specific protein
Proteins are synthesised on ribosomes.
carrier molecule brings specific amino acid
ribosome
mRNA
protein chain forming
amino acids

The completed protein chain folds up into its unique shape which relates to its function — e.g. a structural protein like collagen.

Mutations

MUTATIONS — changes to the sequence of DNA bases. They occur continuously.

	Effect
Most mutations	**No effect on the protein.**
Some mutations	**Alter the protein slightly but its shape and function are not affected.**
Very few mutations	**Change the shape of the protein and affect its function. E.g. an enzyme substrate no longer fits or a structural protein loses its strength.**

Mutations in non-coding DNA can alter how genes are expressed.

Reproduction

Asexual and Sexual Reproduction

	Asexual	Sexual
Parents	**One parent**	**Two parents**
Cell division	**Mitosis**	**Meiosis and mitosis**
Produces	**Genetically identical offspring**	**Offspring containing a mixture of the parents' genes**
Advantages	**Fast compared to sexual reproduction, so many identical offspring can be produced in favourable conditions.** **Only one parent needed so no energy is wasted finding a mate.**	**The variation produced in the offspring increases the chance that some individuals of a species will survive a change in the environment.** **We can use selective breeding to utilise the variation in offspring and increase food production.**

Some organisms can reproduce by both methods depending on the circumstances. E.g. the malaria parasite reproduces sexually in mosquitoes and asexually in humans.

Meiosis

Meiosis produces cells with half the normal number of chromosomes.

The cell duplicates its genetic information.

The cell divides and each new cell has one copy of each chromosome.

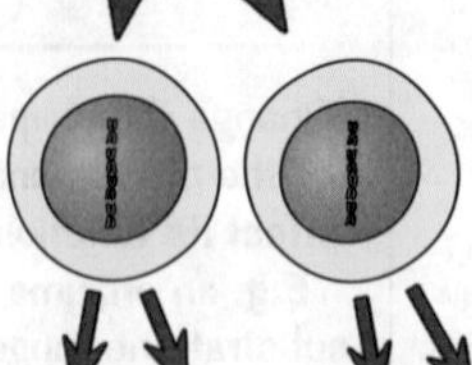

Both cells divide again to make four gametes.

Each gamete only has a single set of chromosomes.

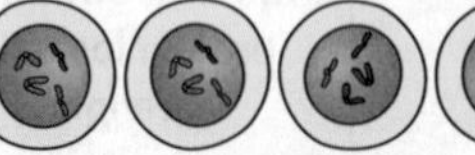

All the gametes produced by meiosis are genetically different.

Gametes

Gametes are formed by meiosis in the reproductive organs.

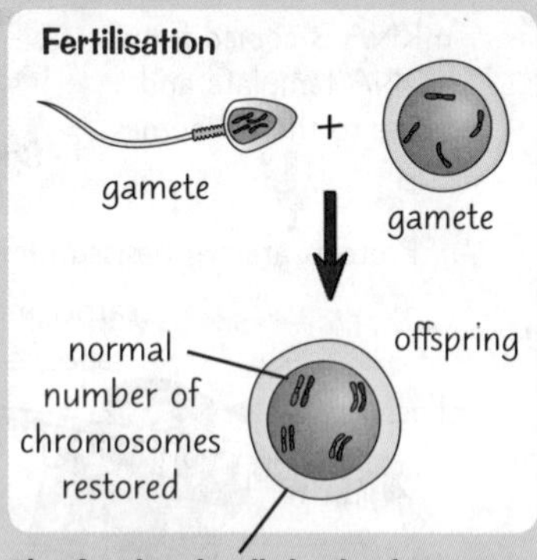
Fertilisation
+
gamete
gamete
offspring
normal number of chromosomes restored

The fertilised cell divides by mitosis many times to form an embryo.

	Animal	Plant
M	Sperm	Pollen
F	Egg	Egg

Genetic Diagrams

Genetic Terms

ALLELE	A version of a gene.
DOMINANT	An allele that is always expressed.
RECESSIVE	An allele that is only expressed when two copies are present.
HOMOZYGOUS	Both of an organism's alleles for a trait are the same.
HETEROZYGOUS	An organism's alleles for a trait are different.
GENOTYPE	An organism's combination of alleles.
PHENOTYPE	The characteristics an organism has.

Two Types of Genetic Diagrams

1 A Punnett square for X and Y chromosome sex determination.

female gametes

	X	X
male gametes X	XX female	XX female
male gametes Y	XY male	XY male

Gametes' genotypes

Offspring's genotypes

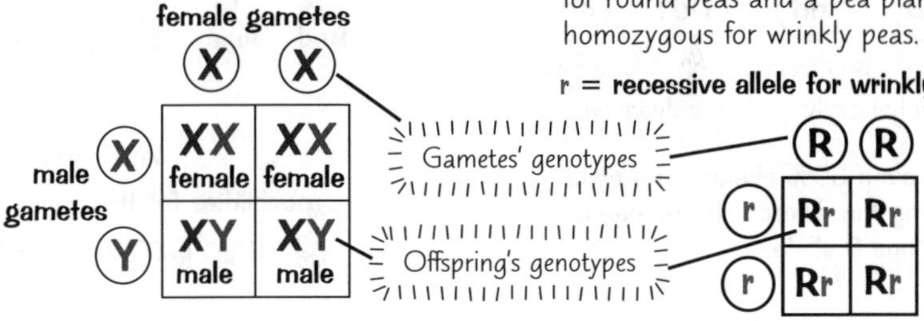

That's a 1:1 ratio of male to female offspring.

A Punnett square for a genetic cross between a pea plant homozygous for round peas and a pea plant homozygous for wrinkly peas.

r = recessive allele for wrinkly peas

	R	R
r	Rr	Rr
r	Rr	Rr

All of the offspring have round peas.

2 A genetic cross between two pea plants that are heterozygous for pea type.

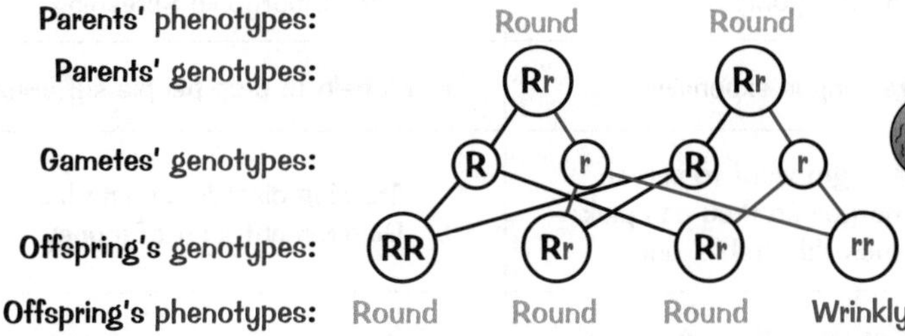

Parents' phenotypes: Round, Round

Parents' genotypes: Rr, Rr

Gametes' genotypes: R, r, R, r

Offspring's genotypes: RR, Rr, Rr, rr

Offspring's phenotypes: Round, Round, Round, Wrinkly

That's a 3:1 ratio of round to wrinkly offspring.

I'm not old, I was born like this!

This example shows a trait with single gene inheritance, like fur colour in mice and red-green colour blindness, but most traits are the result of multiple genes.

Inherited Disorders

Polydactyly

INHERITED DISORDERS — disorders caused by certain alleles that are inherited from the parents.

POLYDACTYLY — a genetic disorder where a baby's born with extra fingers or toes. It's caused by a dominant allele.

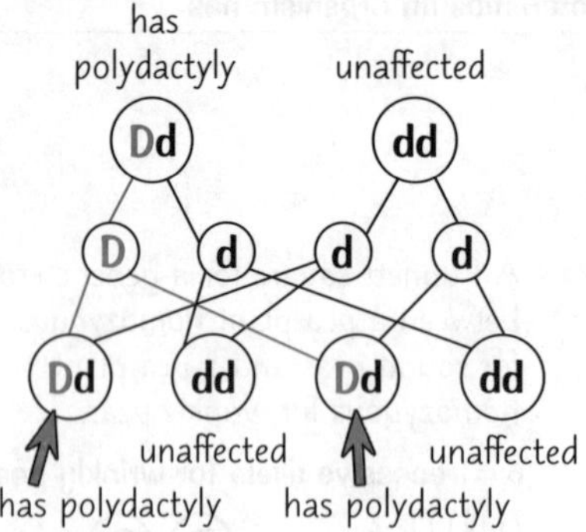

has polydactyly
unaffected
Dd
dd
D
d
d
d
Dd
dd
Dd
dd
unaffected
unaffected
has polydactyly
has polydactyly

There's a 50% chance of a child having the disorder if one parent has one D allele.

Cystic Fibrosis

CYSTIC FIBROSIS — a genetic disorder of the cell membranes. It's caused by a recessive allele.

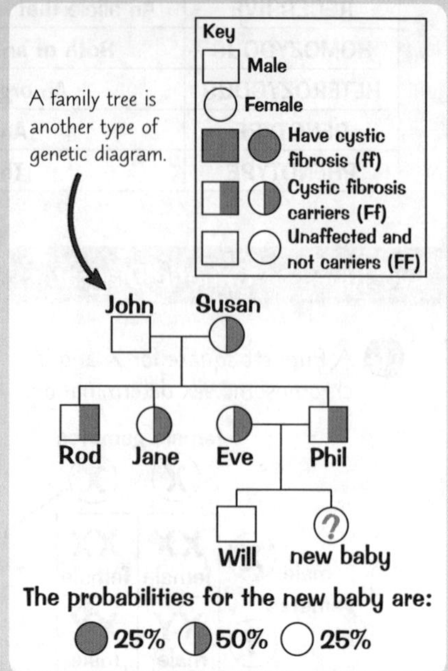

Key
Male
Female
Have cystic fibrosis (ff)
Cystic fibrosis carriers (Ff)
Unaffected and not carriers (FF)
A family tree is another type of genetic diagram.
John
Susan
Rod
Jane
Eve
Phil
Will
new baby

The probabilities for the new baby are:

25% 50% 25%

Embryonic Screening

Against Embryonic Screening	For Embryonic Screening
Screening is expensive.	It will help to stop people suffering.
People might want to screen their embryos so they can pick the most 'desirable' one.	Treating disorders costs the Government a lot of money.
It implies that people with genetic problems are 'undesirable'.	There are laws to stop it going too far.

Developments in Genetics & Variation

Developments in Genetics

Year

1850 — Mendel conducted breeding experiments with plants.

— Mendel published his research.

— Scientists became familiar with chromosomes. They were able to observe how they behaved during cell division.

1900 — Scientists realised there were similarities in the way Mendel's 'units' and chromosomes acted. It was proposed that the 'units' (genes) were found on the chromosomes.

1950 — Structure of DNA determined. Scientists went on to find out exactly how genes work.

Three of Mendel's Conclusions

1. Hereditary 'units' determine characteristics.
2. The units are passed on to offspring unchanged.
3. The units can be dominant or recessive.

Scientists of Mendel's day didn't have the background to properly understand his findings.

Told ya

Variation

VARIATION — differences in the characteristics of organisms.

Genetic variation

Differences in the genes individuals inherit cause variation within a population. This variation is usually extensive.

Mutations cause these differences in genes.

E.g. eye colour

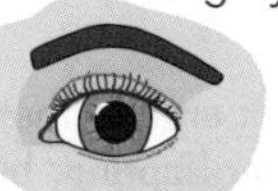

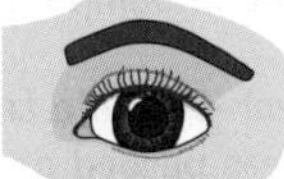

	Most mutations	Some mutations	Very few mutations
Effect on phenotype	None	Slight	New phenotype

Environmental variation

Differences in the conditions in which an organism develops cause variation.

E.g. leaf colour

Genetic and environmental variation

For most characteristics, variation is caused by both genetics and the environment.

E.g. plant height

Evolution and Speciation

Charles Darwin

Best known for: the theory of evolution by natural selection.

Published: On the Origin of Species (1859)

Controversy: the theory challenged some religious ideas.

Research: a world expedition, experiments, discussions, knowledge of geology and fossils.

Setbacks: insufficient evidence to convince many scientists because the mechanism wasn't known about.

Alfred Russel Wallace

Best known for: work on warning colouration in animals and his theory of speciation.

Published: joint writings with Darwin on evolution (1858)

Research: observations from world travels provided evidence for the theory of evolution by natural selection.

Natural Selection

The theory of evolution: **All of today's species have evolved from simple life forms that first started to develop over three billion years ago.**

Species show wide variation.

Limited resources mean organisms are in competition.

Organisms with the most suitable characteristics for the environment are more likely to survive.

These organisms are more likely to breed.

The beneficial characteristics are passed on and gradually become more common in the population.

Jean-Baptiste Lamarck argued that changes acquired during an organism's lifetime were passed on to its offspring, but experiments didn't support this hypothesis.

Speciation

SPECIATION — the development of a new species by natural selection.

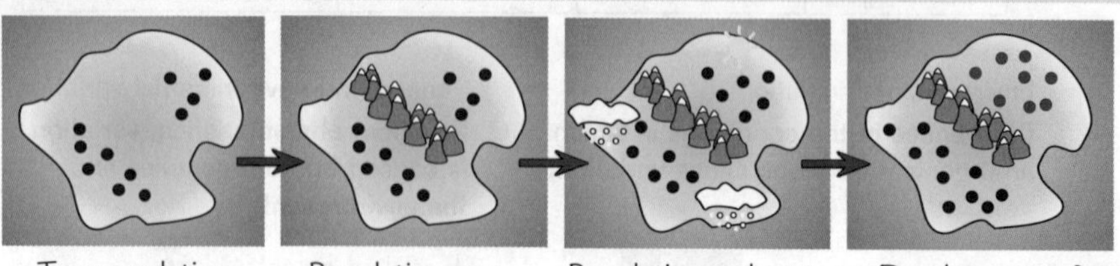

Two populations of a species. → Populations are isolated. → Populations adapt to new environments. → Development of new species.

Two groups become separate species when individuals have changed so much that they won't be able to breed with one another to produce fertile offspring.

Uses of Genetics

Selective Breeding

SELECTIVE BREEDING — **breeding plants or animals for particular characteristics.**

Four uses of selective breeding:

1. **Greater meat or milk production.**
2. **Big or unusual flowers.**
3. **A good, gentle temperament in dogs.**
4. **Disease resistance in crops.**

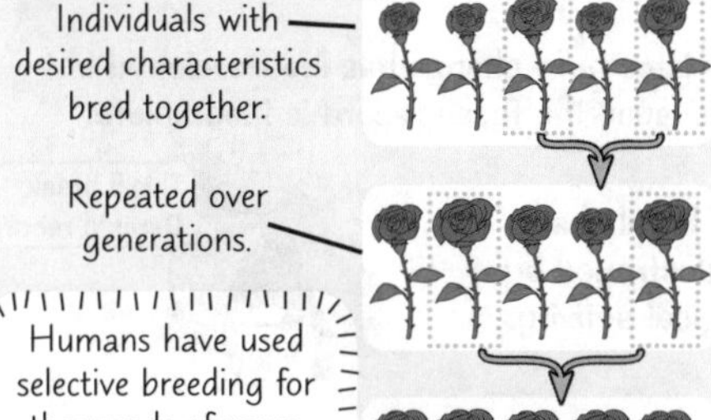
Individuals with desired characteristics bred together.
Repeated over generations.

Humans have used selective breeding for thousands of years. However, it can lead to inbreeding.

Four Types of Cloning

1. **Embryo transplants**
2. **Tissue culture**
3. **Cuttings**

Used to clone plants.

4. **Adult cell cloning**

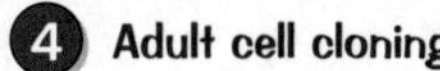

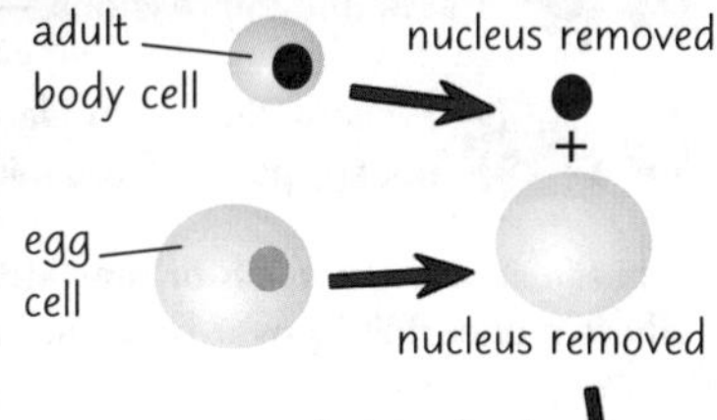
adult body cell
nucleus removed
egg cell
nucleus removed

electric shock makes it divide

live animal (with same genes as adult body cell)

embryo implanted into surrogate mother

Genetic Engineering

Genetic engineering transfers a gene responsible for a desirable characteristic from one organism's genome into another organism. Three uses of genetic engineering:

1. **Genes for producing human insulin transferred to bacteria.**
2. **Genes for bigger and better quality fruit transferred to crops.**
3. **Genes for resistance to insects, diseases, herbicides transferred to crops.**

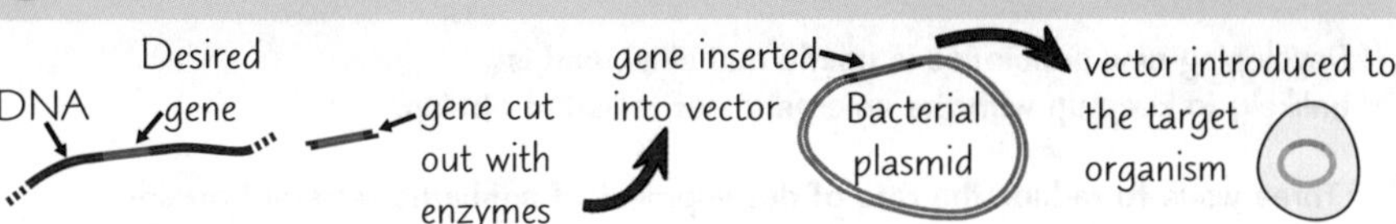
Desired gene
DNA
gene cut out with enzymes
gene inserted into vector
Bacterial plasmid
vector introduced to the target organism

Pros of GM crops	Cons of GM crops
Greater yields	Could reduce biodiversity
Helps people with diets that lack nutrients	Concerns about effects on human health
Already being grown without problems	Transplanted genes may spread into the wild

Fossils and Antibiotic Resistance

Fossils

FOSSILS — remains of organisms from many thousands of years ago. Three ways that fossils form:

Studying fossils and antibiotic-resistant bacteria contributed to the theory of evolution by natural selection being widely accepted.

1.  Gradual replacement by minerals — **happens to slow decaying parts.**
2.  Casts and impressions — **e.g. footprints, burrows and rootlet traces.**
3.  Preservation — **in places where conditions prevent decay, parts of organisms can be preserved.** E.g. in amber, glaciers and peat bogs.

Fossils show how much or how little organisms have changed as life developed but there's uncertainty over how life began because the fossil record is incomplete:

Many early forms of life were soft-bodied and decayed away completely.

Some fossils have been destroyed by geological activity.

This'll break Derek's record.

Antibiotic-Resistant Bacteria

Bacteria can evolve quickly because they reproduce rapidly.

MRSA is a type of antibiotic-resistant bacterium.

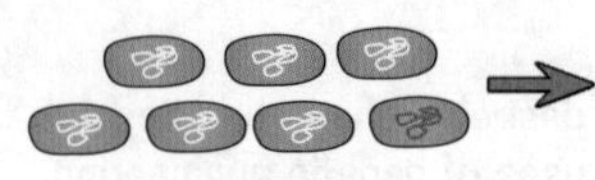

Random mutations can lead to resistance to an antibiotic. → Resistant bacteria survive the antibiotic and reproduce. → Populations of resistant bacteria arise and spread easily, because people aren't immune and there isn't effective treatment.

Developing new antibiotics is costly and slow, and is unlikely to keep up with the emergence of resistant strains.

Three ways to reduce the rate of development of antibiotic-resistant strains:

1.  **Not prescribing antibiotics for non-serious or viral infections.**
2.  **Reducing the use of antibiotics in agriculture.**
3.  **Patients taking the full course of antibiotics to kill all the bacteria.**

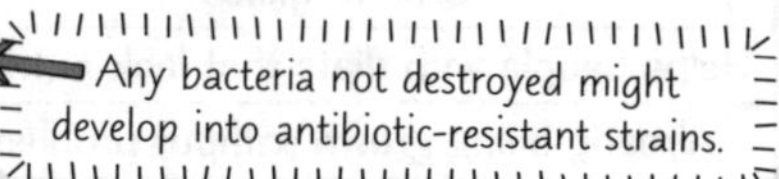

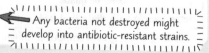

Any bacteria not destroyed might develop into antibiotic-resistant strains.

Classification and Extinction

The Linnaean System

The Linnaean system was developed by Carl Linnaeus.

It organises organisms by their characteristics and structures into these groups.

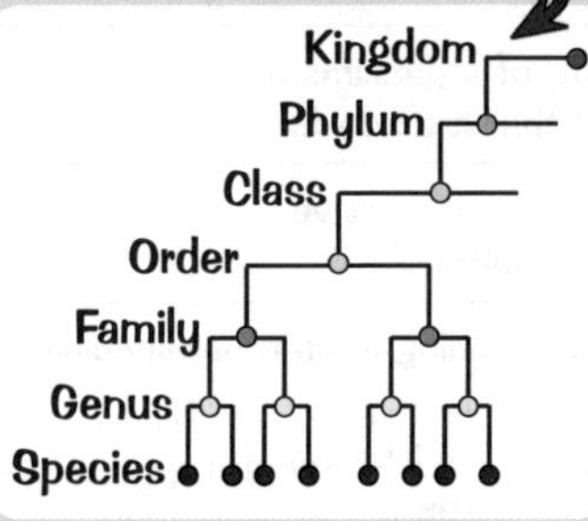
Kingdom
Phylum
Class
Order
Family
Genus
Species

Binomial naming system

	Genus	Species
E.g.	*Homo*	*sapiens*

Developments in Classification

Better microscopes showed us more about internal structures of organisms.

Improved chemical analysis increased our understanding of biochemical processes.

New models of classification proposed, e.g. Carl Woese's three domain system.

Bacteria	true bacteria
Archaea	a different type of prokaryotic cell usually found in extreme places
Eukaryota	including protists, fungi, plants and animals

Evolutionary Trees

Evolutionary trees show how scientists think different species are related to each other.

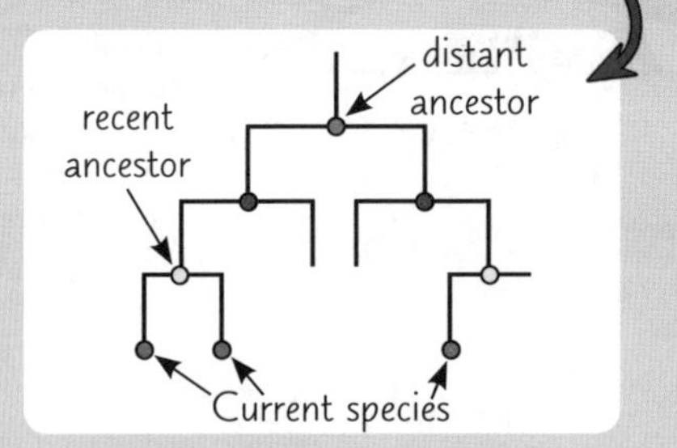
distant ancestor
recent ancestor
Current species

	Living species	Extinct species
Data source	Current classification data	Fossil data

Extinction

EXTINCTION — when no individuals of a species remain.

Environmental change

New predators

New diseases

Competition

Catastrophic events

Basics of Ecology

Definitions of Ecological Terms

POPULATION	All the organisms of one species living in a habitat.
COMMUNITY	The populations of different species living in a habitat.
STABLE COMMUNITY	A community in which all species and environmental factors are in balance, so the population sizes are roughly constant.
ECOSYSTEM	The interaction of a community of organisms and the parts of their environment that are non-living.
ADAPTATION	A feature that enables an organism to survive in the conditions of its normal habitat.
INTERDEPENDENCE	Each species in a community depending upon other species for things, e.g. pollination, food, shelter or seed dispersal.

Due to interdependence, change in an ecosystem (e.g. a species being removed) can affect the whole community.

Factors Affecting Communities

Both biotic (living) and abiotic (non-living) factors can affect organisms in a community:

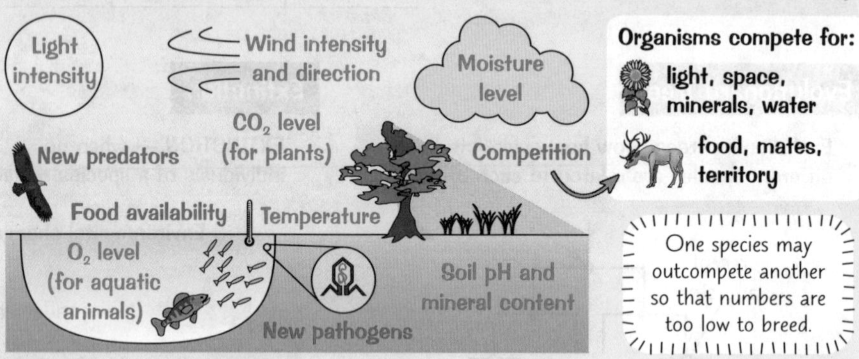
Light intensity
Wind intensity and direction
Moisture level
Organisms compete for:
light, space, minerals, water
food, mates, territory
CO_2 level (for plants)
New predators
Competition
Food availability
Temperature
O_2 level (for aquatic animals)
Soil pH and mineral content
New pathogens

One species may outcompete another so that numbers are too low to breed.

Three Types of Adaptation

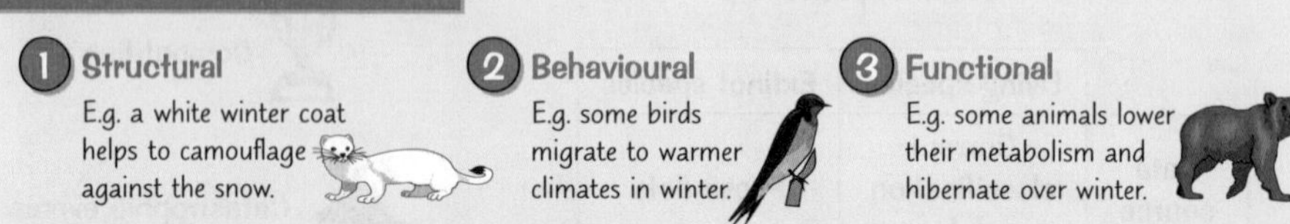
1 Structural
E.g. a white winter coat helps to camouflage against the snow.
2 Behavioural
E.g. some birds migrate to warmer climates in winter.
3 Functional
E.g. some animals lower their metabolism and hibernate over winter.

EXTREMOPHILES — organisms that are adapted to live in extreme conditions, such as high temperature, high pressure or high salt concentration (e.g. bacteria in deep sea vents).

Food Chains & Environmental Change

Food Chains

 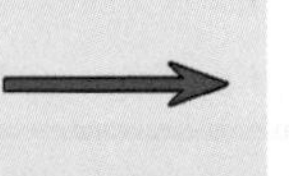   

PRODUCER — a plant or alga that makes glucose by photosynthesis. All food chains start with a producer.

PRIMARY CONSUMER — an animal that eats producers and may be eaten by secondary consumers.

SECONDARY CONSUMER — an animal that eats primary consumers and may be eaten by tertiary consumers.

PREDATOR — a consumer that kills and eats other animals (prey).

BIOMASS — the mass of living material.

Energy stored in biomass is transferred along food chains and used by other organisms to build biomass.

Predator-Prey Cycles

In a stable community the numbers of predators and prey rise and fall in cycles:

Prey population increases

Predator population increases

Prey population decreases
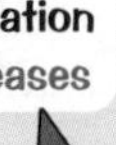
Predator population decreases

Predator-prey cycles are always out of phase with each other, as it takes a while for one population to respond to changes in the other population.

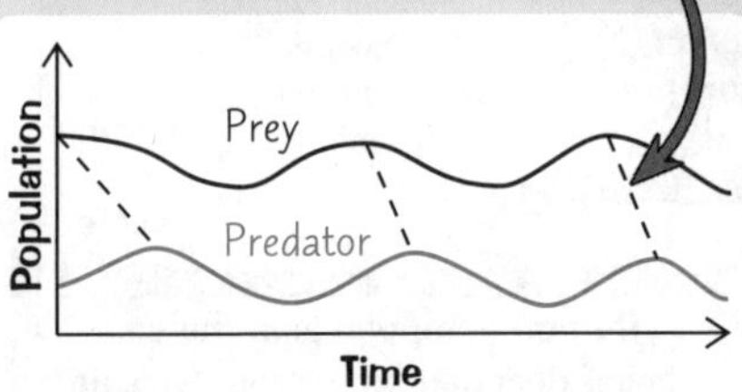
Population
Prey
Predator
Time

Environmental Changes

Environmental changes can cause the distribution of organisms to change.

Three examples of these environmental changes are:

1. **Water availability** — e.g. how much rainfall there is.
2.  **Temperature**
3.  **Atmospheric gases** — e.g. how much air pollution there is.

Changes can be caused by seasonal factors, geographic factors or human interaction.

Cycling of Materials

Recycling Materials

Materials are cycled through the abiotic and biotic parts of an ecosystem.

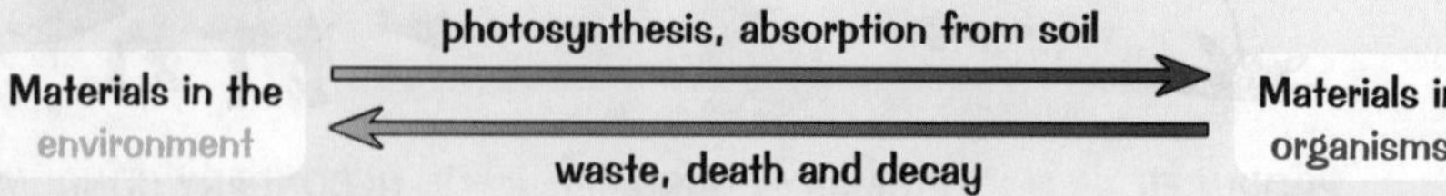
photosynthesis, absorption from soil
Materials in the environment
Materials in organisms
waste, death and decay

Materials decay because they're broken down by microorganisms.
Decay puts materials like mineral ions back into the soil.

The Water Cycle

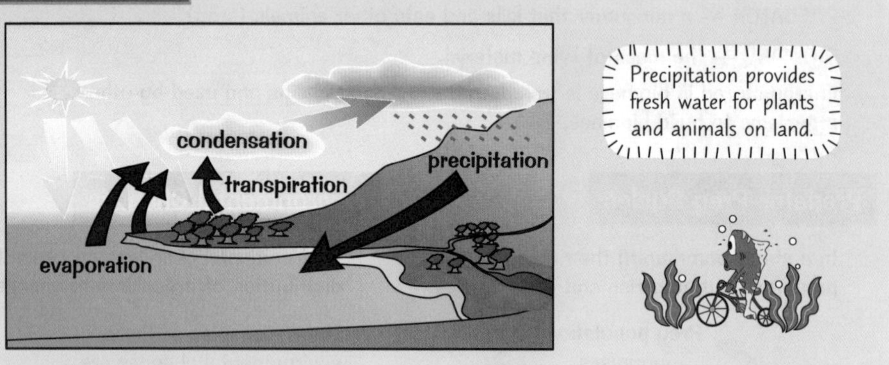
condensation
precipitation
transpiration
evaporation
Precipitation provides fresh water for plants and animals on land.

The Carbon Cycle

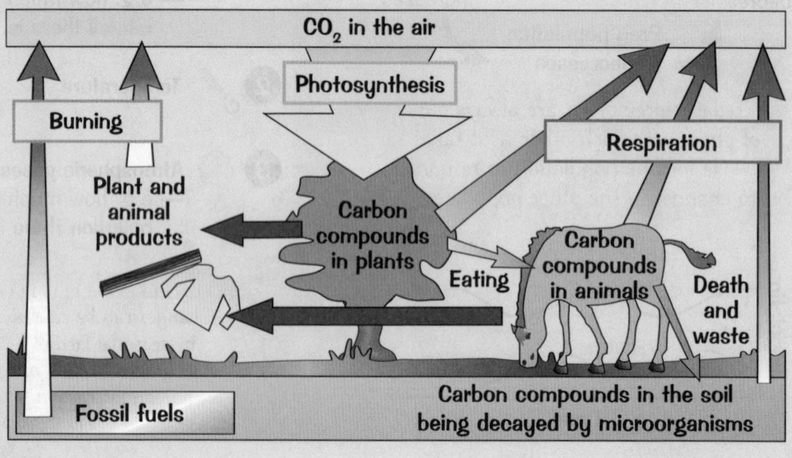
CO_2 in the air
Photosynthesis
Burning
Respiration
Plant and animal products
Carbon compounds in plants
Eating
Carbon compounds in animals
Death and waste
Fossil fuels
Carbon compounds in the soil being decayed by microorganisms

Decay and Biodiversity

Three Factors that Increase the Rate of Decay

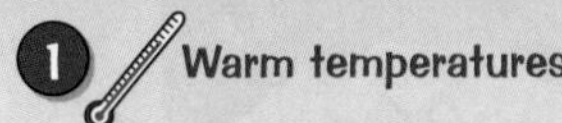
1 Warm temperatures

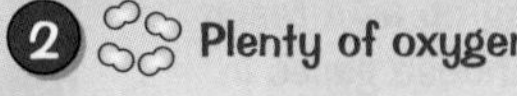
2 Plenty of oxygen

3 Moist conditions

Farmers and gardeners provide ideal conditions for decay when making compost (decomposed organic matter that's used as a natural fertiliser).

These three things create ideal conditions for most organisms involved in decay.

Biogas

Biogas is produced by anaerobic decay (decay without oxygen). It's mainly made up of methane, which can be burned as a fuel.

Biogas is made on a large scale from waste material (e.g. sludge from sewage works) in biogas generators.

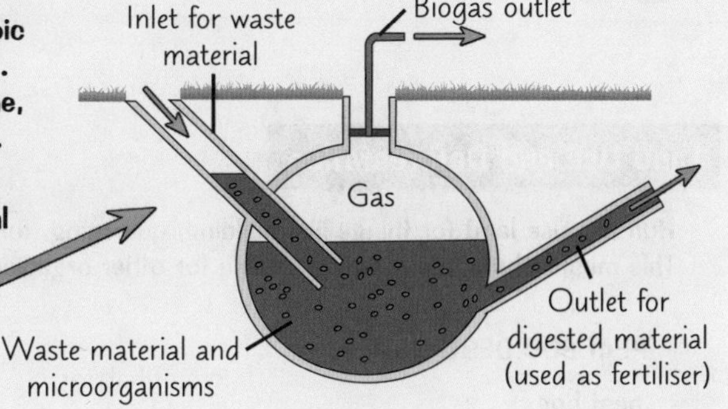

Inlet for waste material
Biogas outlet
Gas
Waste material and microorganisms
Outlet for digested material (used as fertiliser)

Biodiversity

BIODIVERSITY — the variety of different species on Earth, or within an ecosystem.

HIGH BIODIVERSITY (lots of different species)

Reduced dependence of one species on another for things like food, shelter and the maintenance of the physical environment.

STABLE ECOSYSTEMS

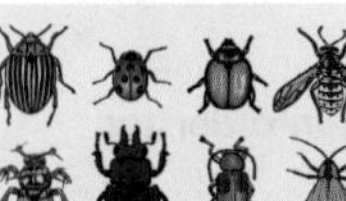

For the human species to survive, it's important that a good level of biodiversity is maintained.

Sadly, human actions are reducing biodiversity — we've only recently started taking measures to stop this.

Human Effects on Ecosystems

Global Warming

The Earth is gradually heating up as a result of increasing levels of greenhouse gases in the atmosphere.

radiation trapped

greenhouse gases (e.g. carbon dioxide and methane)

Three consequences of global warming could be:

1.  Rising sea levels (so low-lying places may flood).
2.  A change in the distribution of some organisms.
3.  A decrease in biodiversity (as some species may become extinct).

Land Use and Deforestation

Humans use land for things like building, quarrying, farming and dumping waste. This means there's less land available for other organisms.

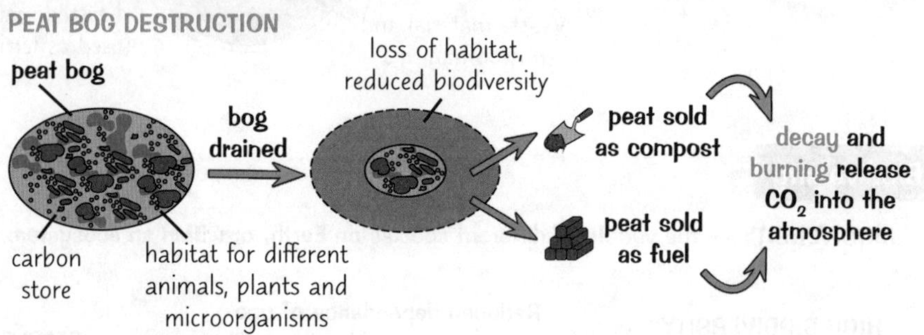
PEAT BOG DESTRUCTION
peat bog
carbon store
habitat for different animals, plants and microorganisms
bog drained
loss of habitat, reduced biodiversity
peat sold as compost
peat sold as fuel
decay and burning release CO_2 into the atmosphere

DEFORESTATION — the cutting down of forests.

 It has been done on a large-scale in tropical areas in order to:

 clear land for rice fields and farming cattle,

 grow crops to make biofuels.

Maintaining Ecosystems

Pollution

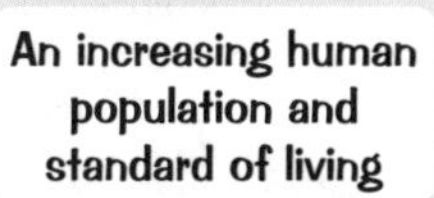

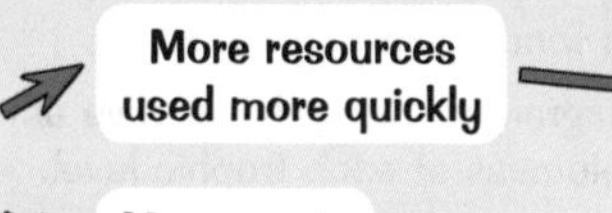

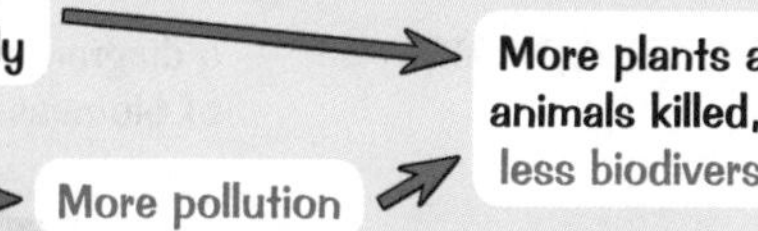

An increasing human population and standard of living → More resources used more quickly → More plants and animals killed, so less biodiversity

An increasing human population and standard of living → More waste produced → More pollution → More plants and animals killed, so less biodiversity

These are three ways we pollute the environment:

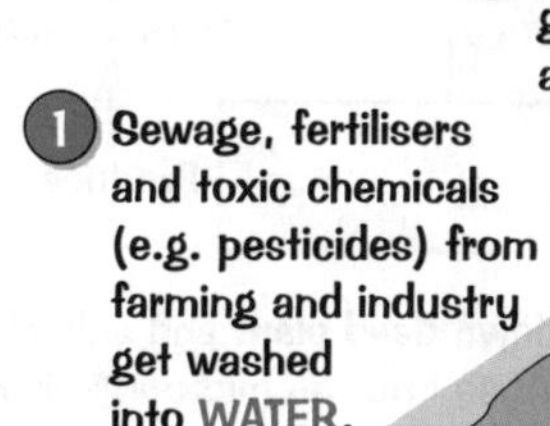

1. Sewage, fertilisers and toxic chemicals (e.g. pesticides) from farming and industry get washed into WATER.
2. Smoke and acidic gases released into the atmosphere pollute AIR.
3. Toxic chemicals (e.g. from farming) and waste dumped in landfill sites pollute LAND.

Waste has to be handled properly to reduce pollution of the environment.

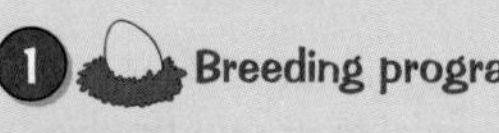

Five Programmes to Protect Ecosystems

1. **Breeding programmes** — endangered species are bred in captivity to make sure the species survive.

2. **Habitat restoration** — rare habitats like mangroves, heathland and coral reefs are protected and regenerated.

3. **Hedgerows and field margins** — these are reintroduced around fields where only a single crop type is grown, creating habitats for more organisms.

4. **Government regulations** — e.g. to reduce deforestation and CO_2 emissions.

5. **Recycling** — reduces the amount of waste going to landfill sites.

Programmes like these are set up to reduce the negative effects of human activities on ecosystems and biodiversity.

Trophic Levels

Trophic Levels and Biomass

TROPHIC LEVEL — a stage in a food chain.

PYRAMID OF BIOMASS — a diagram showing the relative amounts of biomass at each trophic level.

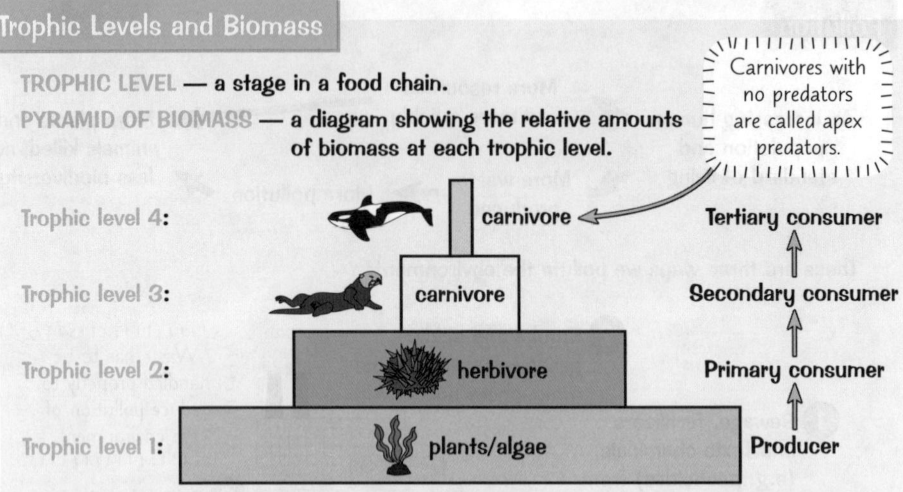
Carnivores with no predators are called apex predators.
Trophic level 4:
carnivore
Tertiary consumer
Trophic level 3:
carnivore
Secondary consumer
Trophic level 2:
herbivore
Primary consumer
Trophic level 1:
plants/algae
Producer

Decomposers (e.g. bacteria) secrete enzymes to break down dead plant and animal matter into small soluble food molecules. These then diffuse into the microorganisms.

Biomass Transfer

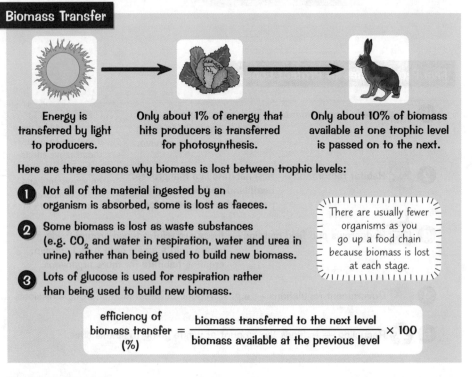

Energy is transferred by light to producers. → **Only about 1% of energy that hits producers is transferred for photosynthesis.** → **Only about 10% of biomass available at one trophic level is passed on to the next.**

Here are three reasons why biomass is lost between trophic levels:

1. **Not all of the material ingested by an organism is absorbed, some is lost as faeces.**
2. **Some biomass is lost as waste substances (e.g. CO_2 and water in respiration, water and urea in urine) rather than being used to build new biomass.**
3. **Lots of glucose is used for respiration rather than being used to build new biomass.**

There are usually fewer organisms as you go up a food chain because biomass is lost at each stage.

$$\text{efficiency of biomass transfer (\%)} = \frac{\text{biomass transferred to the next level}}{\text{biomass available at the previous level}} \times 100$$

Food Security and Biotechnology

Food Security Threats

FOOD SECURITY — having enough food to feed a population. It's threatened by these six things:

1.  Increasing birth rate
2.  Changing diets in developed countries (meaning they take scarce food resources from other countries).
3.  New pests and pathogens that affect crops and livestock.
4.  Environmental changes that affect farming (e.g. changes in rainfall patterns).
5.  Cost of farming
6.  Conflict

Sustainable methods of food production are needed to feed everyone now and in the future.

Increasing Efficiency

Food can be produced more efficiently by reducing the energy transferred from livestock to the environment, for example:

 by restricting movement.

 by keeping animals in temperature-controlled environments.

High-protein food can also be fed to animals to increase growth.

Overfishing

Fish stocks are declining due to overfishing. We need to maintain stocks at a level where the fish continue to breed. This can be done by:

 Introducing fishing quotas

 Controlling net size

Biotechnology

BIOTECHNOLOGY — where living things and biological processes are used and manipulated to produce a useful product. It can be used to produce food for the growing human population (e.g. microorganisms can be cultured for use as a food source).

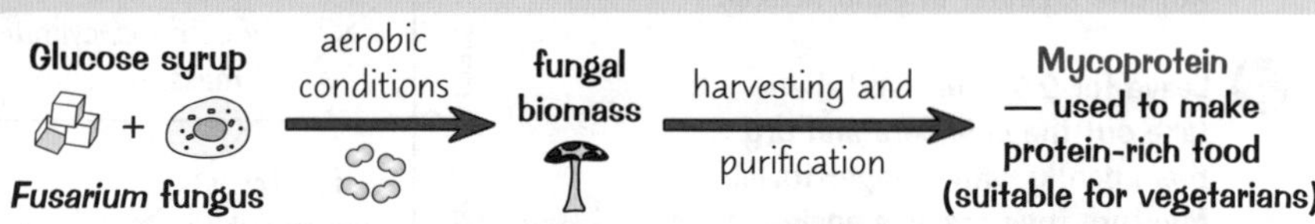

Glucose syrup
+
Fusarium fungus
aerobic conditions
fungal biomass
harvesting and purification
Mycoprotein — used to make protein-rich food (suitable for vegetarians)

Genetically modified crops can produce more food or food with a greater nutritional value — e.g. 'golden rice' produces a chemical that's converted into vitamin A in the body.

Human insulin can be produced by genetically engineered bacteria.

Required Practicals 1

Microscopy

Start with the lowest-powered lens then move the stage up with the coarse adjustment knob.

↓

Look down the eyepiece and adjust the focus with the adjustment knobs (use the coarse one first).

↓

To see the slide with a greater magnification, swap to a higher-powered lens and refocus.

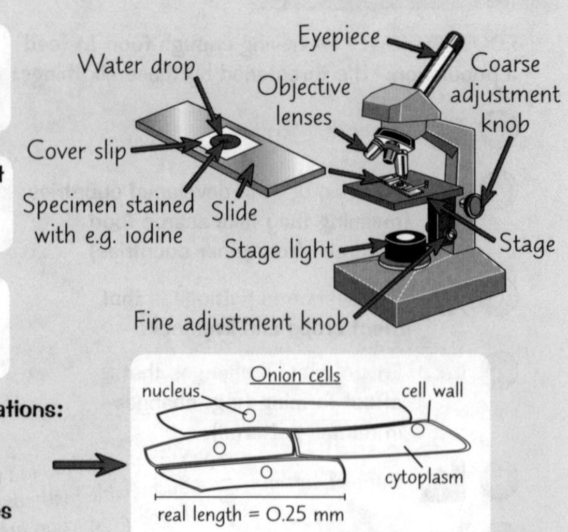
Eyepiece
Water drop
Coarse adjustment knob
Objective lenses
Cover slip
Specimen stained with e.g. iodine
Slide
Stage light
Stage
Fine adjustment knob
Onion cells
nucleus
cell wall
cytoplasm
real length = 0.25 mm
magnification = × 100

When drawing your observations:

- use a sharp pencil
- draw unbroken lines
- label important features
- include a magnification scale.

Osmosis

Four steps to investigate the effect of different concentrations of sugar or salt solutions on plant tissue:

1. Cut a potato into identical cylinders.
2. Measure the mass of each cylinder.
3. Prepare beakers containing different concentrations of sugar or salt solution and one of pure water. Put one cylinder in each beaker.
4. Leave for 24 hours and then take out the cylinders and dry them gently with a paper towel. Measure their masses again.

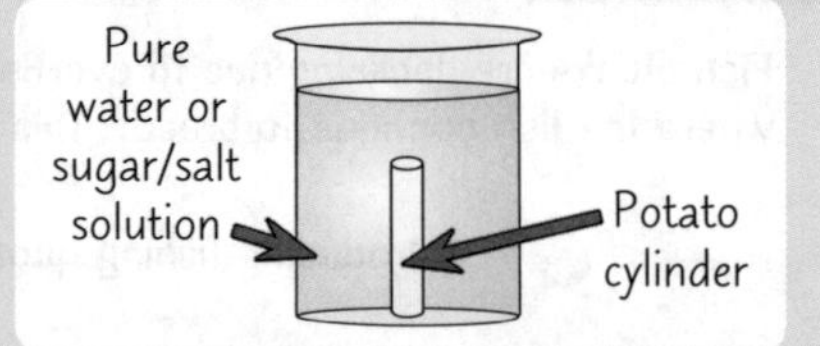
Pure water or sugar/salt solution
Potato cylinder

Independent Variable	Dependent Variable
concentration of sugar/salt solution	potato cylinder mass

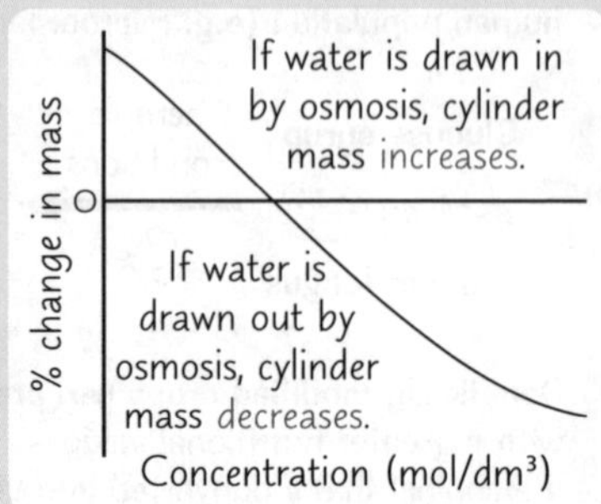
% change in mass
0
If water is drawn in by osmosis, cylinder mass increases.
If water is drawn out by osmosis, cylinder mass decreases.
Concentration (mol/dm³)

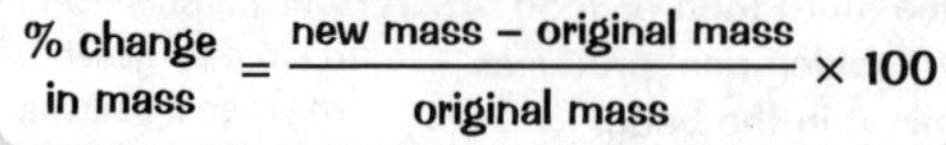

$$\text{\% change in mass} = \frac{\text{new mass} - \text{original mass}}{\text{original mass}} \times 100$$

Required Practicals 2

Antibiotics and Bacterial Growth

Five steps to prepare an uncontaminated culture of bacteria in a lab:

1. **A Petri dish and some agar jelly are sterilised (e.g. heated to a high temperature) to kill any unwanted microorganisms in them.**
2. **Hot agar jelly is poured into the Petri dish and allowed to cool and set.**
3. **An inoculating loop is passed through a hot flame to sterilise it.**
4. **The inoculating loop is used to transfer bacteria to the agar jelly.**

Petri dish
agar jelly

5. **The Petri dish is lightly taped shut to stop microorganisms in the air getting in. It is stored at 25 °C and upside down to stop drops of condensation falling on the agar surface.**

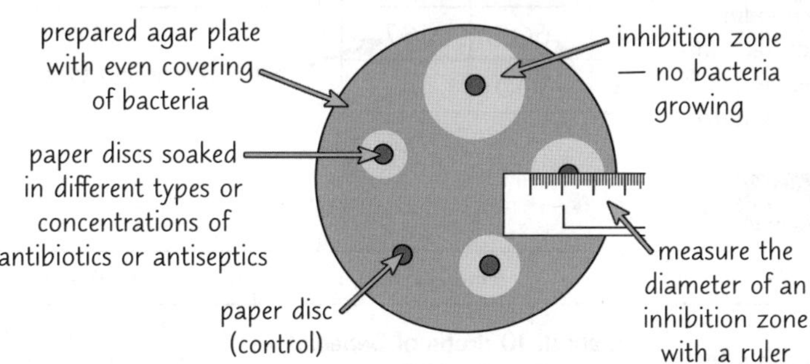
prepared agar plate with even covering of bacteria
inhibition zone — no bacteria growing
paper discs soaked in different types or concentrations of antibiotics or antiseptics
measure the diameter of an inhibition zone with a ruler
paper disc (control)

Independent Variable	Dependent Variable
type or concentration of antibiotic or antiseptic	**width of inhibition zone**

More effective antibiotic/ antiseptic → **Wider inhibition zone**

area of inhibition zone = πr^2

r is the radius (half the diameter)

The equation can also be used to calculate the area of a bacterial colony. You just need to measure the diameter of the colony you're interested in.

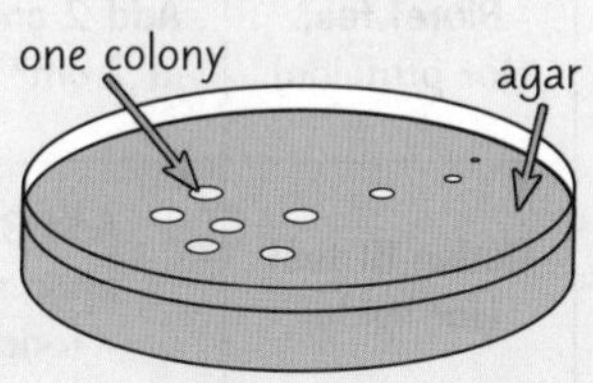
one colony
agar

Required Practicals 3

Effect of pH on Amylase Activity

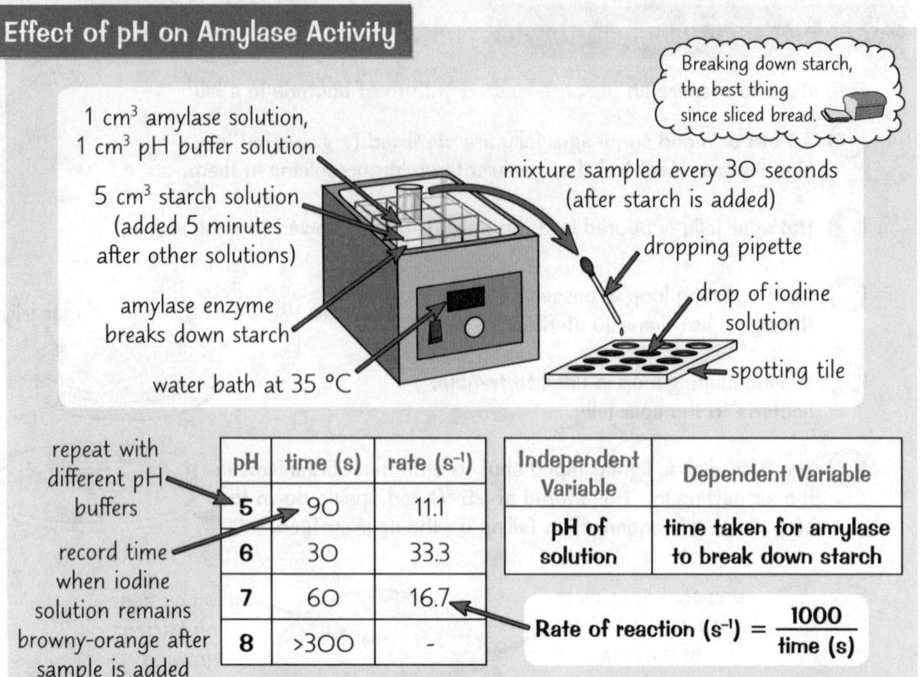

Breaking down starch, the best thing since sliced bread.
1 cm³ amylase solution, 1 cm³ pH buffer solution
5 cm³ starch solution (added 5 minutes after other solutions)
amylase enzyme breaks down starch
water bath at 35 °C
mixture sampled every 30 seconds (after starch is added)
dropping pipette
drop of iodine solution
spotting tile
repeat with different pH buffers
record time when iodine solution remains browny-orange after sample is added

pH	time (s)	rate (s^{-1})
5	90	11.1
6	30	33.3
7	60	16.7
8	>300	-

Independent Variable	Dependent Variable
pH of solution	time taken for amylase to break down starch

$$\text{Rate of reaction } (s^{-1}) = \frac{1000}{\text{time (s)}}$$

Food Tests

Test	Method	Positive result
Benedict's test (for sugars)	Add about 10 drops of Benedict's solution to a 5 cm³ food sample and leave for 5 minutes at 75 °C.	blue → green, yellow, brick red
Iodine test (for starch)	Add a few drops of iodine solution to a 5 cm³ food sample and mix.	browny-orange → blue-black
Biuret test (for proteins)	Add 2 cm³ of biuret solution to a 2 cm³ food sample and mix.	blue → purple
Sudan III test (for lipids)	Add 3 drops of Sudan III stain solution to a 5 cm³ food sample and mix.	Mixture separates into two layers — top layer bright red

Required Practicals 4

Effect of Light Intensity on Photosynthesis Rate

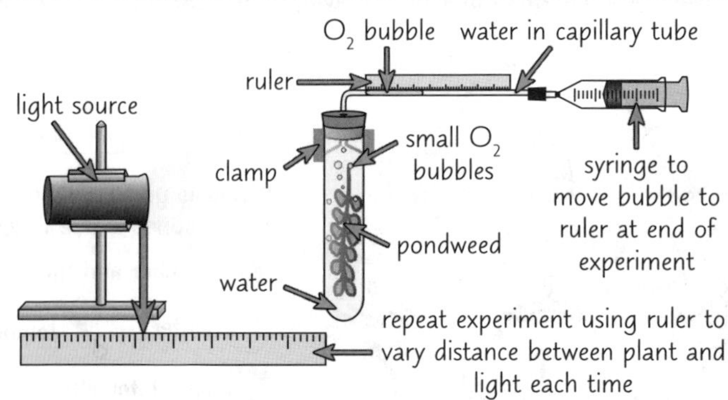
O2 bubble
water in capillary tube
ruler
light source
clamp
small O2 bubbles
syringe to move bubble to ruler at end of experiment
pondweed
water
repeat experiment using ruler to vary distance between plant and light each time

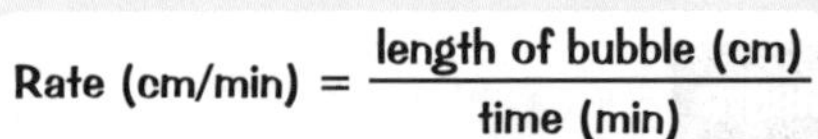

$$\text{Rate (cm/min)} = \frac{\text{length of bubble (cm)}}{\text{time (min)}}$$

Independent Variable	Dependent Variable
distance from light	length of bubble

Higher light intensity
Faster rate of photosynthesis
Faster rate of O2 production

Reaction Time

REACTION TIME — the time it takes to respond to a stimulus.

Four steps to investigate reaction time:

1. **Hold a ruler between the thumb and forefinger of the person being tested.**
2. **Drop the ruler without warning and record the distance it falls before it's caught.**
3. **Repeat the test several times then calculate the mean distance that the ruler fell.**
4. **Repeat the experiment to investigate the effect of a factor on reaction time.**
   E.g. give the person being tested a caffeinated drink, wait ten minutes and then repeat the experiment.

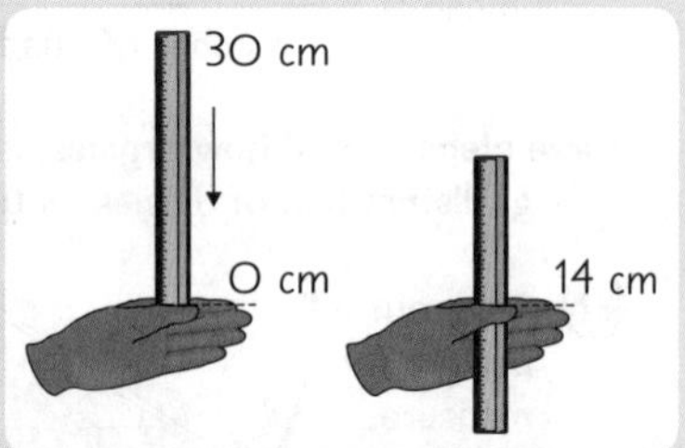
30 cm
0 cm
14 cm

The further the ruler falls, the slower the reaction time.

Required Practicals 5

Plant Growth Responses

You can investigate the effect of light or gravity on the growth of seedlings, e.g.:

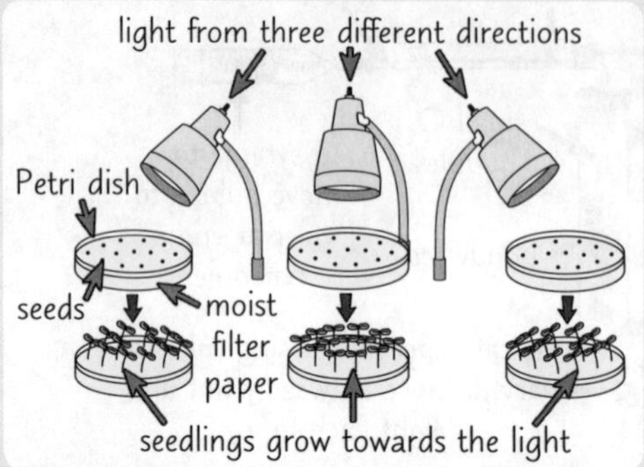
light from three different directions
Petri dish
seeds
moist filter paper
seedlings grow towards the light

Independent Variable	Dependent Variable
direction of light	direction of growth

Variables you should control to know the response is due to light only:

- number and type of seeds
- water
- temperature
- light intensity

Distribution and Abundance of Organisms

Four steps to estimate population size of small organisms using quadrats:

1. **Place a 1 m^2 quadrat at random in a field.**
2. **Count all the daisies within it.**
3. **Repeat several times and work out the mean number of daisies per quadrat.**

$$\text{mean} = \frac{\text{total number of organisms}}{\text{number of quadrats}}$$

E.g. 154 daisies in 7 quadrats means there are 22 daisies per m^2.

4. **Multiply the size of the entire area by the number of organisms per m^2.**

E.g. if the field has an area of 800 m^2, the population size estimate is 17 600 daisies.

Watch out for wild quad-rats.

Three steps to find how organisms are distributed across an area using transects — e.g. distribution of daisies as you move away from the edge of a pond:

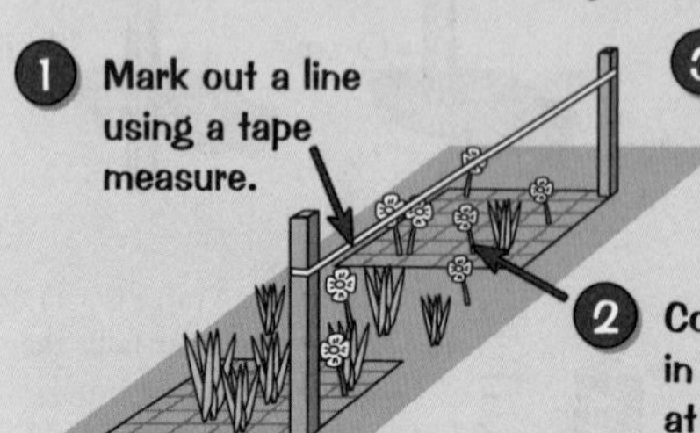

1. **Mark out a line using a tape measure.**
2. **Count the daisies in quadrats placed at regular intervals along the line.**
3. **Draw a graph to show how the daisies are distributed.**

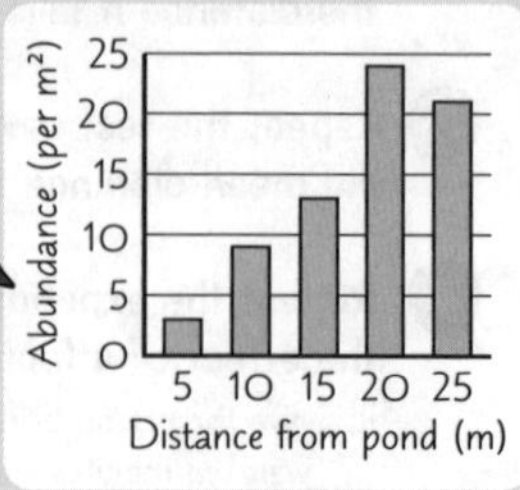
Abundance (per m²)
25
20
15
10
5
0
5
10
15
20
25
Distance from pond (m)

Required Practicals 6

Effect of Temperature on Rate of Decay

Six steps to investigate the effect of temperature on the rate of decay of fresh milk:

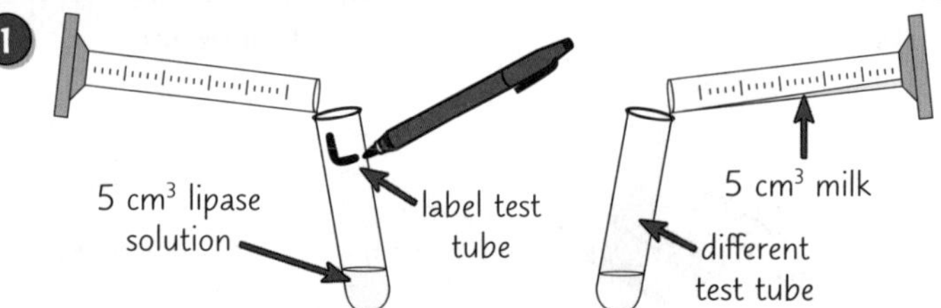
1
5 cm3 lipase solution
label test tube
5 cm3 milk
different test tube

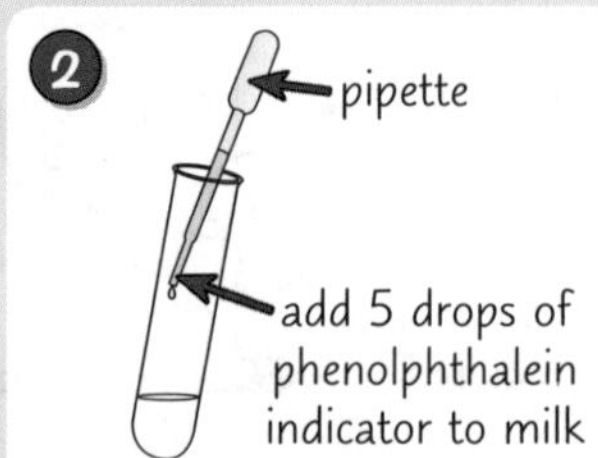
2
pipette
add 5 drops of phenolphthalein indicator to milk

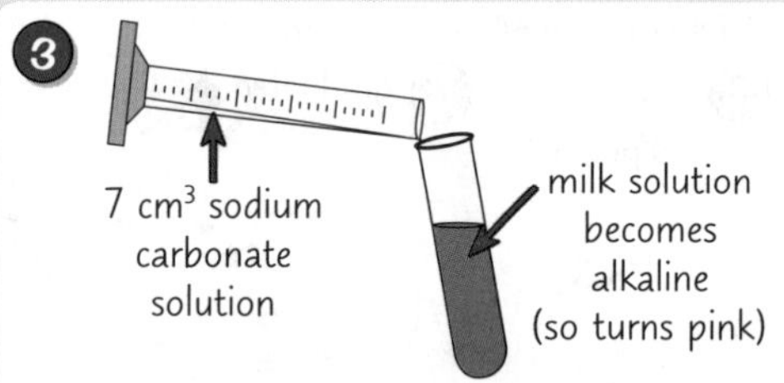
3
7 cm3 sodium carbonate solution
milk solution becomes alkaline (so turns pink)

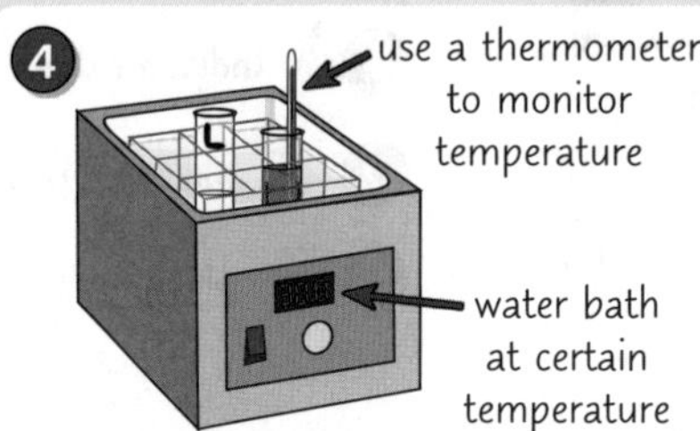
4
use a thermometer to monitor temperature
water bath at certain temperature

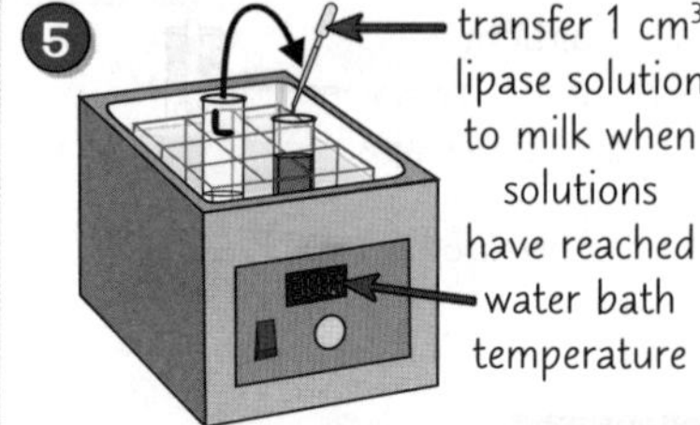
5
transfer 1 cm3 lipase solution to milk when solutions have reached water bath temperature

6
start a stopwatch immediately
00:01

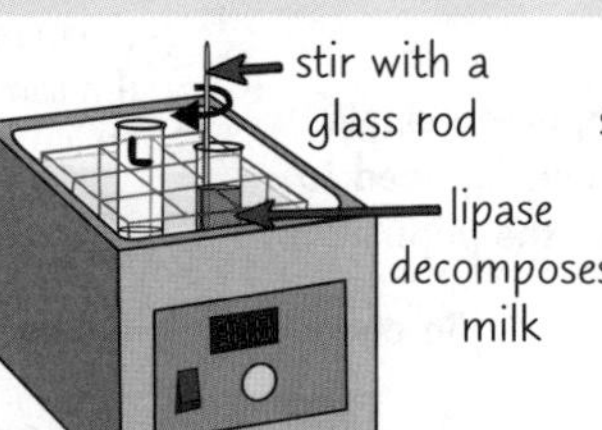
stir with a glass rod
lipase decomposes milk

milk broken down, so no longer alkaline

colour change from pink to white

stop the stopwatch
00:24

Rate of decay (s^{-1})

Temperature (°C)

repeat several times at different temperatures

calculate and plot the mean rate at each temperature

Independent Variable	Dependent Variable
temperature	**time taken for colour change**

Measuring and Sampling

Taking Measurements

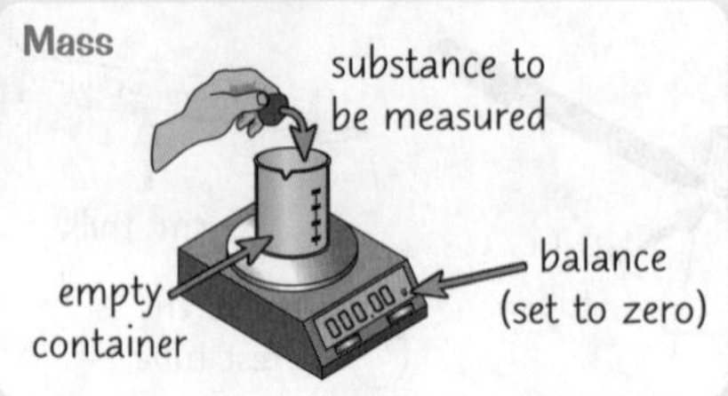

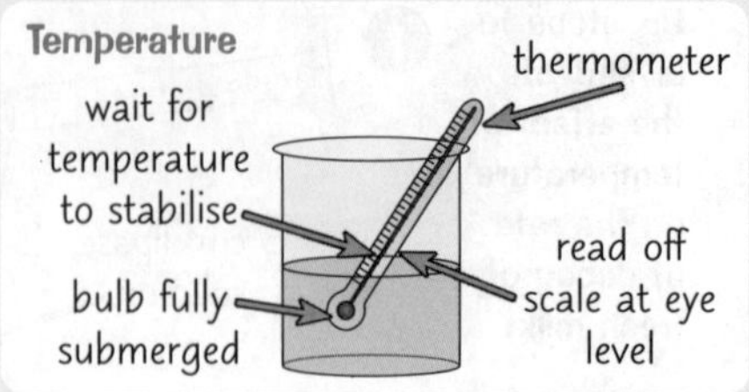

Mass

substance to be measured

empty container

balance (set to zero)

Temperature

thermometer

wait for temperature to stabilise

read off scale at eye level

bulb fully submerged

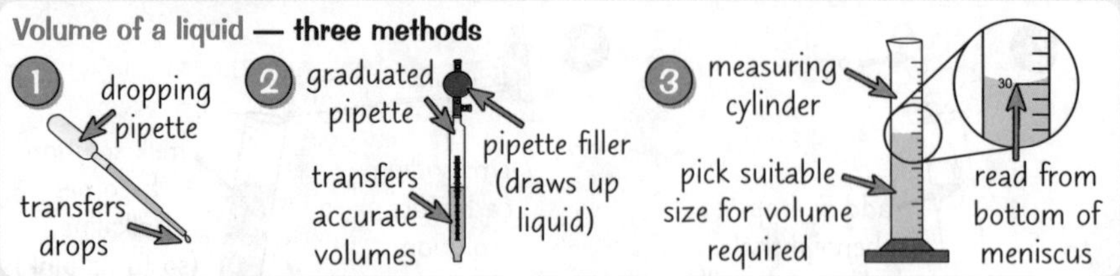

Volume of a liquid — three methods

1. dropping pipette — transfers drops
2. graduated pipette — pipette filler (draws up liquid) — transfers accurate volumes
3. measuring cylinder — pick suitable size for volume required — read from bottom of meniscus

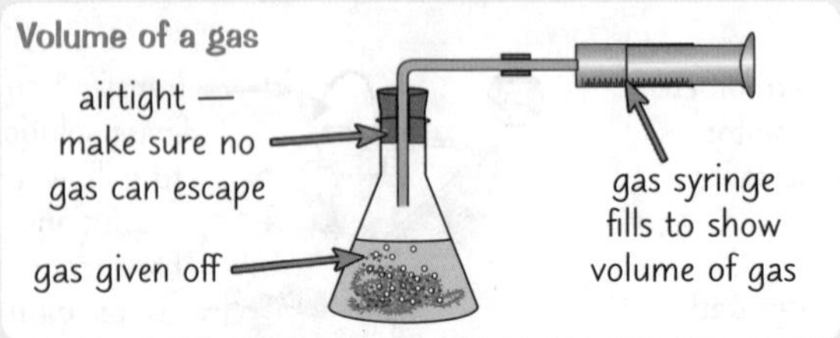

Volume of a gas

airtight — make sure no gas can escape

gas given off

gas syringe fills to show volume of gas

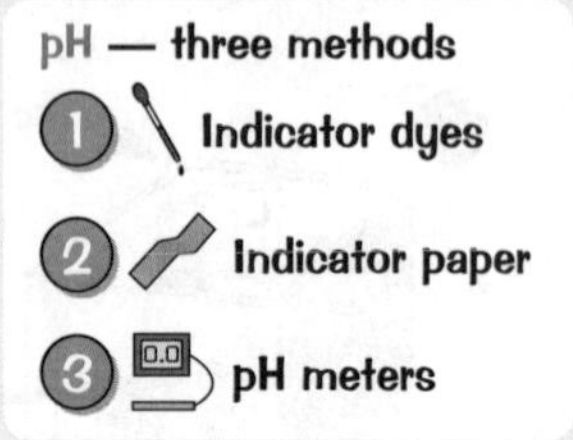

pH — three methods

1. Indicator dyes
2. Indicator paper
3. pH meters

Sampling

It's generally not possible to study every single organism in a population, so you usually need to take a random sample to represent the population.

If a sample doesn't represent the population as a whole, it's said to be biased.

Assign a number to each member of the population.

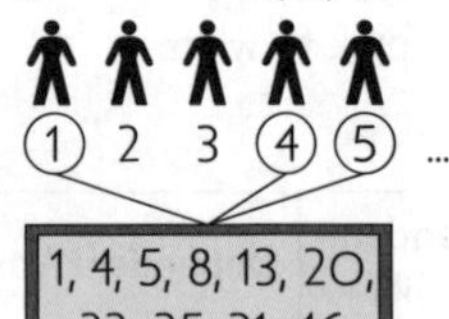
1 2 3 4 5 ...
1, 4, 5, 8, 13, 20, 22, 25, 31, 46

Use a random number generator to choose the sample group.

To choose sample sites from a large area:

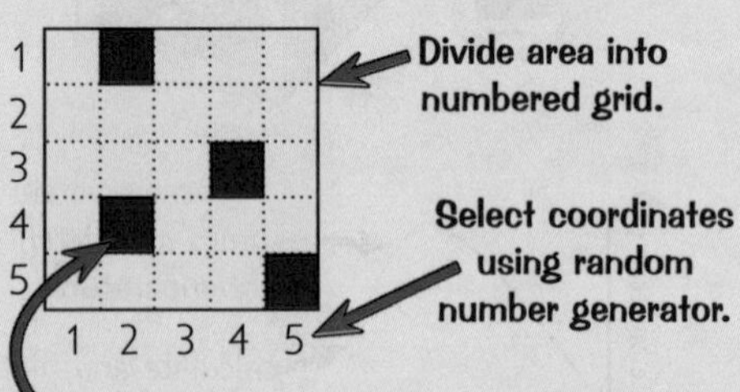
1
2
3
4
5
1 2 3 4 5
Divide area into numbered grid.
Select coordinates using random number generator.

Take samples at chosen coordinates (e.g. using quadrats or transects).

Practical Techniques

Heating

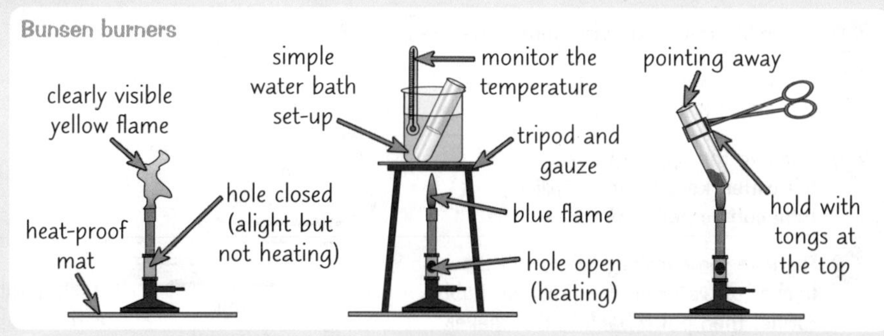
Bunsen burners
clearly visible yellow flame
heat-proof mat
hole closed (alight but not heating)
simple water bath set-up
monitor the temperature
tripod and gauze
blue flame
hole open (heating)
pointing away
hold with tongs at the top

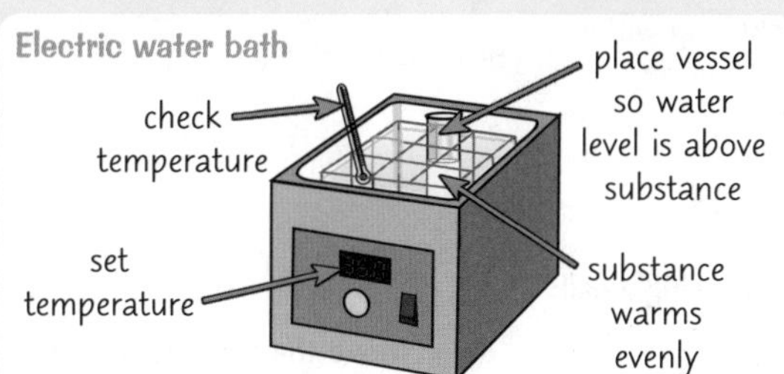
Electric water bath
check temperature
set temperature
place vessel so water level is above substance
substance warms evenly

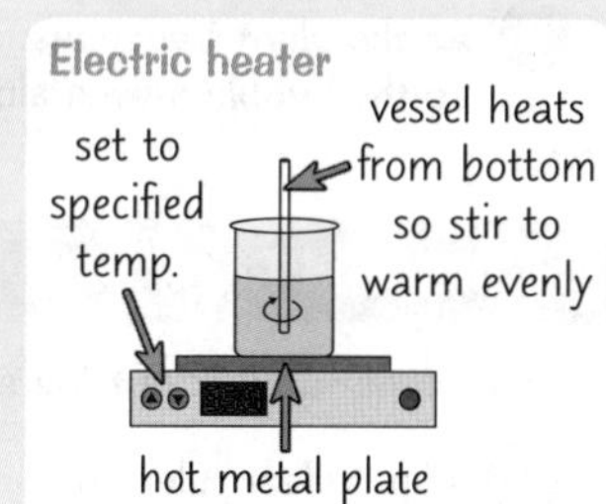
Electric heater
set to specified temp.
vessel heats from bottom so stir to warm evenly
hot metal plate

Safety

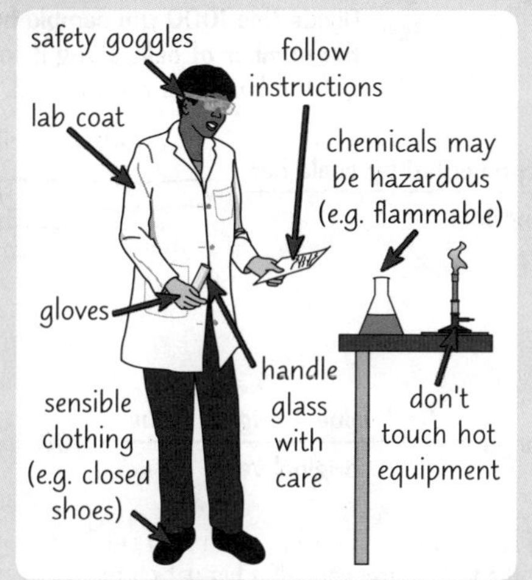
safety goggles
follow instructions
lab coat
chemicals may be hazardous (e.g. flammable)
gloves
handle glass with care
don't touch hot equipment
sensible clothing (e.g. closed shoes)

Ethics

Any organisms used in an investigation need to be treated safely and ethically.

Animals kept in the lab should be cared for in a humane way.

Wild animals captured for study should be returned to their original habitat.

Other students shouldn't be forced or pressured into participating in experiments.

Techniques and Results

Five Steps to Measure Rate of Water Uptake

1. Assemble the potometer underwater.
2. Cut the shoot at a slant underwater and insert it in the potometer.
3. Remove the apparatus from the water, keeping the capillary tube submerged in the beaker.
4. Remove the capillary tube from the beaker of water until a single air bubble forms, then put it back in the beaker.
5. As the plant transpires, it takes up water, so the bubble moves along the scale.

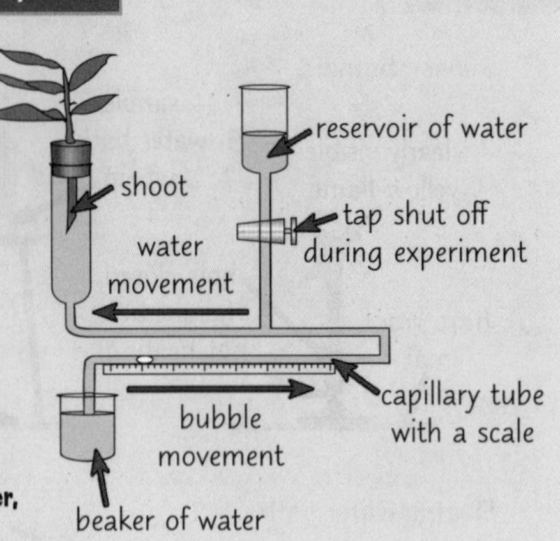
reservoir of water
shoot
tap shut off during experiment
water movement
capillary tube with a scale
bubble movement
beaker of water

Cell Size and Scale Bars

Four steps to work out the size of a single cell:

1. Clip a clear ruler on top of your microscope slide onto the stage.
2. Pick the objective lens that gives an overall magnification of × 100 and focus the image.
3. Move the ruler so that the cells are lined up along 1 mm.

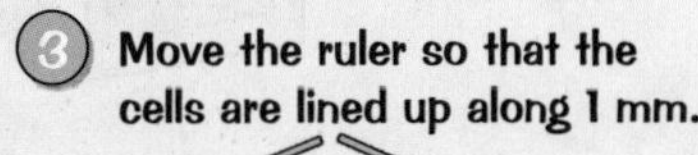

4. Divide this 1000 μm sample by the number of cells along it to get the length of a single cell.

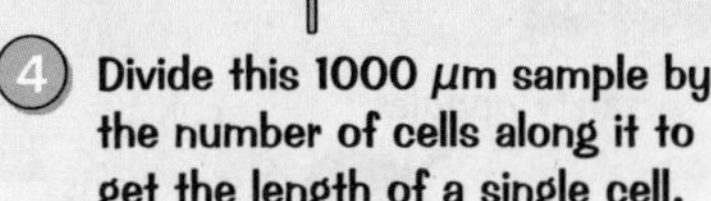

$$\text{scale bar length} = \frac{\text{drawn cell length} \times \text{length represented by scale bar}}{\text{actual cell length}}$$

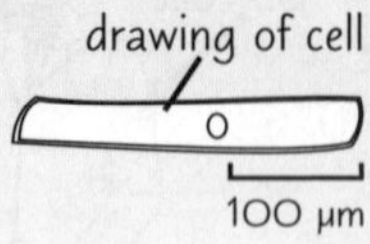
drawing of cell
100 μm

Percentage Change

To compare results that didn't have the same initial value, calculate the percentage change:

$$\%\ \text{change} = \frac{\text{final value} - \text{original value}}{\text{original value}} \times 100$$

Hooray! You did it — you got through ALL the facts. Give yourself a big pat on the back.